Truth-Science

A journey like none other,
Into the realm of Truth

Copyrights

TRUTH-SCIENCE: A journey like none other, into the realm of Truth.

Title ID: 3540808
ISBN-13: 978-0615436517
ISBN-10: 061543651X
Such a Time

Truth-Science

A journey like none other,
Into the realm of Truth

Contents

Part II: Truth from Biblical Perspectives

Part III: Truth-Science Methodology

Contents

Part II: Truth from Biblical Perspectives

Part III: Truth-Science Methodology

Part IV: Truth-Science Philosophy

TO THE ETERNAL, TRUE WITNESS,
WHO OBSERVES ALL THINGS
FROM ALL PERSPECTIVES

Chapter One
Introduction to Truth-Science

When you live in a mega-product, mega-information age, you struggle to keep up with everything: the foods you eat, the beverages you drink, the medicines you take, the vehicles you drive, the clothing you wear, the water you use to cook with and bathe in, the air you breathe, and so many other things that keep you wondering if you're really being told the truth about the safety of the many products and environments you come into contact with each and every day.

Lengthy warnings, disclaimers, and fine print, backed by one or more scientific studies, accompany most products that are consumed. You are constantly being bombarded by new and often conflicting information about the products you consume, stirred up perhaps by the competition. Yet the science behind a product is so important because if a product has not been proven to be safe or superior by one or more scientific studies or standards, consumers are generally hesitant to make such purchases.

Science, whether it is social, physical or psychological, was once so highly esteemed in our society that it was often revered as a god, influencing both individuals as well as groups. There was a time in America during the nineteen fifties and sixties when scientifically backed products were rarely ever questioned by the general public, but they were accepted as absolute fact by most consumers when advertised on the major television networks or radio media that once held the trust of the American public.

However, more and more individuals have become skeptical about the science, because it has become so politicized along with the mainstream media that disseminates much of the latest information. So many people are confused as to why there is such contradiction coming out of the science sectors. One study says one thing, while another later study contradicts the former. So what *should* we believe? What is the truth?

Truth in science is so essential. And just as good science helped to stop the tyranny of bad religion in past centuries (to be discussed later), scientific proof and actual truth must become as synonymous as possible to stop the tyranny of science now gone political. Science is best driven by the honest quest for truth, rather than by the special interests of the business or political structures, but unfortunately, the big business and mainstream-political structures seem to be in control since they have the economic power whip to drive the horses wherever they desire.

Furthermore, it is difficult to know who is really looking out for you, and who has your best interests at heart. There does not seem to be an ethical scientific oath that scientists universally swear by, like the Hippocratic Oath for doctors. There is only a prevailing scientific method, but even this is not as pure and untainted as you might think. Those who desire unbiased truth should carefully scrutinize this methodology in order to expunge it of its Progressive political influences.

Although *Truth-Science* looks at some of the Progressive influences that have affected modern day science, *Truth-Science* is mostly about truth itself. It examines truth from both human and Biblical perspectives and it identifies what truth *is* in its most simplified form, a subject most often attempted by answering the question, "What is *the* truth," rather than "What *is* truth."

Truth-Science is absolutely not an extension of *Natural Science.* In fact, many people are naïve as to what the term Natural Science really means. Most people generally associate the term *natural* to mean of the earth, untainted or uncontaminated, and basically a good thing that is considered pure. However, it does not mean that at all, and since the classical sciences that make up today's mainstream sciences generally fall under the common classification of Natural Science, it is essential that the true meaning of this term be made very clear.

So, what are Natural Sciences? It is important to understand that at their core, the *Natural Sciences*[1] all have a covert code of ethics, a foundational philosophy that is based on the doctrine of *Naturalism*. Naturalism is defined as action, inclination or thought based *only* on natural desires and instincts. It is a theory that denies that objects or events have any supernatural significance, and it is the doctrine that believes all phenomena can be explained and accounted for by using *only* scientific laws. *Naturalism*[2] is therefore an Atheistic

---

[1] "natural science." Merriam-Webster Online Dictionary. 2010. Merriam-Webster Online. 18 May 2010 <http://www.merriam-webster.com/dictionary/natural science>

[2] "naturalism." Merriam-Webster Online Dictionary. 2010. Merriam-Webster Online. 18 May 2010 <http://www.merriam-webster.com/dictionary/naturalism>

doctrine. *Humanism*[3] has also been incorporated into the sciences as well. Both Naturalism (circa 1641) and Humanism (1832) sprang from *Atheism*[4] (1546).

No God	*Man is God*	*No Miracles*
Atheism **(root)** (1546) A disbelief in the existence of deity. The doctrine that there is no deity.	**Naturalism** **(branch)** (Circa 1641) An action, inclination or thought based only on natural instincts and desires. A theory denying that an event or object has a supernatural significance; *specifically* : the doctrine that scientific laws are adequate to account for all phenomena.	**Humanism** **(branch)** (1832) A doctrine, attitude, or way of life centered on human interests or values; *especially* : a philosophy that usually rejects supernaturalism and stresses an individual's dignity and worth and capacity for self-realization through reason.

Included within the doctrines of Atheism, Naturalism and Humanism are postulates and precepts that are taught

[3] "humanism." Merriam-Webster Online Dictionary. 2010. Merriam-Webster Online. 18 May 2010 <http://www.merriam-webster.com/dictionary/humanism>

[4] "atheism." Merriam-Webster Online Dictionary. 2010. Merriam-Webster Online. 18 May 2010 <http://www.merriam-webster.com/dictionary/atheism>

regularly in the United States Government education system. All mainstream sciences have been affected by these doctrines. They have produced generations of students, many of whom are now adults, including myself, and who have all been programmed to varying degrees into the atheistic, naturalistic and humanistic ways of thinking.

To better understand the effects of these doctrines in US Government education and within the sciences in particular, the philosophy behind the scientific method should not only be scrutinized more carefully, but more importantly, philosophy's own foundation should first be considered.

Truth-Science looks at philosophy's foundation from the perspective of what is relevant and most important regarding what truth actually is in its most simplified form. This is not in any way insignificant, for one's own personal understanding and agreement of what truth *is* becomes the most influencing variable of that individual's philosophical thinking, affecting the perceptions and the entire belief system of that individual.

One's understanding of truth, and one's relationship with truth, becomes the foundation that supports the myriad of thought structures erected within one's own heart and mind. These thought structures influence what an individual does and does not do. If these thought structures are not founded or supported on truth, then the very actions and deeds of that individual may be found lacking significance in the end.

The Bible says, "For no one can lay any foundation other than the one already laid, which is Jesus Christ."[5] *Truth-Science* explores *why* that is. It does so both Biblically and deductively. The Bible also teaches that if you hold to the truth, you shall know the truth, and the truth shall set you free.[6] When you know what truth really is, your life may never be the same. You may be very surprised by the amazing things that can happen when the truth sets you free!

The term *science*[7] literally means the state of knowing. *Truth-Science* therefore is the state of knowing truth. Most of the contention between religion and science has been caused primarily because each side has been gathering facts in support of their religious-political or atheistic-political agendas. Moreover, it is because people on both sides of the fence have been blindly accepting theories, sophisms, false doctrines, and hypotheses as truth when, in fact, they are not true at all, or they are not entirely true.

---

[5] 1 Corinthians 3:11 NIV

[6] John 8:31-32

[7] "science." Merriam-Webster Online Dictionary. 2010. Merriam-Webster Online. 1 May 2010 <http://www.merriam-webster.com/dictionary/science>

Aristotle once described sound law and custom in his written work on rhetoric:[8] He said that a judge should not be perverted by a litigant moving him to anger, envy or pity, and that doing so would be like someone warping a carpenter's rule before using it. Aristotle said that a litigant's sole responsibility is clearly to show that the alleged fact is or isn't so, or that it did or did not happen. And whether something is important or not, just or unjust, it is up to the judge to decide, and that the judge should not allow the litigants to dictate the importance or unimportance of the matter as the laws that were given had left undefined.

Considering Aristotle's advice for judges, it may also be wise for us as well to judge information and knowledge by allowing the facts to speak for themselves, and by making every effort not to be misguided by the emotional appeals of those who would use anger, envy, pity or any other diversion in an attempt to steer us from the actual truth of the matter as represented and substantiated by the unbiased facts—by letting truth have its own voice, so to speak. Then—and only then—can we rightly judge how it all fits together.

Truth-Science shows how truth, in its most basic form, can be discernable both deductively as well as Biblically, and

---

[8] http://classics.mit.edu/Aristotle/rhetoric.mb.txt

that truth and God are inseparable. There is no truth without God, and both truth and God remain unaffected by the perceptions and misperceptions of humanity.

It is not as difficult a subject as one might think. When presented with the simple facts about what truth really is, you may wonder why it remains so elusive to so many. It will take you only a short time to read this book. Yet what you learn about truth can benefit you foundationally for the rest of your life. You don't have to be a scientist or a student of science to understand it, or to benefit from it. It only takes an honest and open mind.

Part I: Truth from Human Perspectives

Chapter Two
Descartes' Philosophy

The French philosopher, René Descartes [Pronounced, dA-'kärt] (1596-1650), is often referred to as the Father of Modern Philosophy. While searching for the simplest truth on which to base his new philosophy, and after ridding himself of all doubt, he wrote: "je pense, donc je suis," (French) meaning, "I am thinking, therefore I am;" Later on he wrote the more popular, "Cogito, ergo sum," Latin, meaning, "I think, therefore I am."

René Descartes expressed in his writings that he believed everything he "perceived clearly and precisely was true."[9] He began his new philosophy by rejecting all statements that were arguable as to being true or not. He sought to find the most basic truth, a common ground on which everyone could agree; and this common ground or common agreement on what was true became the foundation and premise of his new philosophy, "an unarguable truth that no one of a right mind could disagree with."

Descartes' philosophy was built on the assumption that truth existed in the first place, a logical assumption that rational people could not deny. Yet, to Descartes and others of his day, it existed only as they perceived it to be, and only as they could agree and acknowledge that it was indeed the truth. It is important to point out here that Descartes' philosophy can be described as a "perceived truth" philosophy, an inverted or reversed philosophy that puts the cart before the horse. In other words, truth is not what it *is*, but rather it is what the perceiver perceives it to be. This seems to be a philosophy centered on self-importance, rather than truth's importance.

---

[9] *The Philosophical Writings of Descartes*, translated by Cottingham, Stoothoff, Murdoch and Kenny, Cambridge UP, 1991.

Modern day philosophy still holds to that same notion, that truth is only truth if it is perceived and acknowledged as such. It seems the general concept of truth is similar to the cliché concept of beauty: *Beauty is in the eye of the beholder.* Believing that beauty is only in the eye of the beholder devalues what beauty is, and it makes the beholder self-important and superior to that which is beautiful. So it stands to reason that if people hold to *that* concept of beauty, then it is probably only skin deep as well for them. The truth is more likely to be as the song goes, "Everything is beautiful in its own way." It simply takes a beholder with a keen enough eye to recognize it. Does beauty exist? Yes. Does everyone recognize beauty when they see it? No.

But don't be so naïve to believe that *everything* is beautiful either – ugly exists too, and so do hatred, murder, deception and cruelty. The point is—many people today pay little or no attention to the logic and philosophies that guide their own personal beliefs. Their life principles come from songs, sayings, television shows, books, newspapers, quotes, advertisements and straight from hell, to name only a few of the sources.

Consider this, a commonly quoted example of the extremities of philosophical thinking within US society regarding truth: "If a tree falls in the middle of the forest, and there is no one there to hear it, does it still make a sound?"

Who hasn't heard that ridiculous question and then repeated it to someone else? It is, of course, a nonsensical question. Certainly it makes noise! The laws of physics don't exist only if someone perceives them. So why would truth be any different? If truth exists, but there is no one there to perceive it and acknowledge it, does truth still exist? Absolutely!

Amazingly, that style of thinking has also become the caliber of political chatter we find being broadcast during each election season by those who are, or who are trying to become our leaders. It is a time when so-called truth about issues and candidates tends to fly in all directions, hitting the fan on or just before Election Day. We believe we are electing candidates who stand for the issues they say they stand for. But we really don't know the truth until after they've served, or when it surprisingly stings us in the pocketbook later on, or when we discover our freedoms or protections have disappeared. Even the accomplishments of the candidates are perceived so oppositely by the rivaling parties.

The notion of what truth *is* remains subjective to this day. A born-again believer might profess, "Jesus is the truth," but that believer generally cannot answer the question *why* is Jesus the truth when asked, other than to say, "Because the Bible says so", or "Because I believe he is."

No doubt, René Descartes believed in truth, but it does not seem that he had an understanding of what truth actually is, apart from his assumption that if all doubt is gone, the only thing left is truth. He understood doubt, and encouraged people in their right minds to get rid of all doubt in order to arrive at the truth, but it appears he stopped there.

Unfortunately, the "perceived truth" philosophy of Descartes tends to pervade western thinking and US Government education today, most particularly in the disciplines of science. It is a subtle philosophy, one that has been carefully integrated into the mainstream sciences. But make no mistake—man is *not* the center of the universe.

Chapter Three
Darwinian Philosophy

Charles Darwin (1809-1882) is considered by many to be the *Father of Evolutionary Biology*. Charles Darwin's first and most popular book, *The Origin of Species*, is a compilation of many valid scientific observations, and it appears that Darwin was a very talented and detailed observer; however, the book unfortunately also includes Darwin's opinions, explanations and assumptions which are not only based on seemingly flawed perspectives and conjectures of how modern day and fossilized plants and animals came into existence, but the book also contains an erroneous premise that has served to replace God as being the Creator of all living things, and its popularity has since propelled the modern day course of biological science philosophically toward atheism by omitting God from science altogether.

It appears, according to Darwin's Autobiography, that the basic literary form for his work, *The Origin of Species,* may have been modeled after Euclid and particularly after two books written by William Paley (1743-1805). From Darwin's Autobiography:

> "In order to pass the B.A. examination, it was also necessary to get up Paley's 'Evidences of Christianity,' and his 'Moral Philosophy.' This was done in a thorough manner, and I am convinced that I could have written out the whole of the 'Evidences' with perfect correctness, but not of course in the clear language of Paley. The logic of this book and, as I may add, of his 'Natural Theology,' gave me as much delight as did Euclid. The careful study of these works, without attempting to learn any part by rote, was the only part of the academical course which, as I then felt and as I still believe, was of the least use to me in the education of my mind. I did not at that time trouble myself about Paley's premises; and taking these on trust, I was charmed and convinced by the long line of argumentation."

It is interesting that Darwin was "charmed and convinced by the long line of argumentation" found in Paley's *Evidences*— [if only Darwin could have written as clearly and plainly as did both William Paley and Euclid! If you have never read any of Darwin's original works, you should; however, it may take you a long period of time to adapt to his style of writing]. Note also that Darwin didn't trouble himself with trying to understand Paley's premises at all, having been "charmed and convinced by the long line of argumentation" alone.

Because Darwin was so intrigued by this style of writing, it makes one wonder if perhaps he may have intentionally incorporated this long argument technique into his own works, believing that by doing so he might better convince his readers, instead of having to rely solely on his main premise. As evidenced in the later chapters of *The Origin of Species*, Darwin admittedly struggled with a couple aspects of his own theory.

The Theory of Evolution contains *some* concepts or premises that are true, and which are indeed valid, and backed by excellent observations. These valid concepts however serve to candy coat the Theory of Evolution pill, making it easier for a gullible person to swallow.

For example, "natural selection" and "the struggle for life," or "survival of the fittest," as it later was called, are parts of the Theory of Evolution that contain valid observations and perceptions, and these are generally what are taught in science classes at elementary through high school levels effectively deceiving the students into believing that the entire Theory of Evolution is sound science.

But the problem is this: This type of teaching makes the subtle link between the true perceptions of *natural selection* and *survival of the fittest*, by connecting them to and incorporating them within the framework of the Theory of Evolution, whereby positive mutations occur over an extended period of time resulting in a higher form of life. And this part of Evolution is nothing more than mere conjecture based on an erroneous premise that is both contrary and inconsistent with the tested laws of physics and with Biblical accounts as well.

Natural selection and survival of the fittest are not theories—they are observable facts; but just because a bird snatches a dark colored gypsy moth resting on a light colored tree more readily than a light colored moth, it doesn't mean nature is improving the species. It just means dark moths on light tree trunks are easier to spot by birds. The design was already complete and encoded within the moth species; So the light colored moths in that area will eventually make up the largest population as long as there are light colored trees in that area, because with fewer dark colored moths, the light colored moths will mate more frequently with other light colored moths, resulting genetically in more light colored offspring. In the area of the forest where there are mostly dark colored tree trunks, the dark colored moths will prevail. This explains why we find clusters of species of plants and animals found in various places all over the world rather than an equal distribution of plant and animal species everywhere. It also demonstrates the interrelationships that exist between plants and animals.

For this reason, only the accurate observations and perceptions such as those already mentioned should be included and remain a part of real science. Those portions of the Theory of Evolution that are flawed analysis and poor conjecture based on false premises should be separated out as mere science fiction, for they certainly have no place being taught as mainstream scientific fact in any reputable form of public or private education.

It should also be pointed out here that Charles Darwin was not a true Naturalist. Philosophically speaking, Naturalism is the doctrine that claims, "The world can be understood without recourse to spiritual or supernatural explanations." That being so, it is interesting that Charles Darwin, who considered himself to be a Naturalist, wrote the following paragraph in *The Origin of Species*, Chapter 14, Recapitulation and Conclusion:

> "Why, if man can by patience select variations most useful to himself, should nature fail in selecting variations useful, under changing conditions of life, to her living products? What limit can be put to this power, acting during long ages and rigidly scrutinising the whole constitution, structure, and habits of each creature, -- favouring the good and rejecting the bad? I can see no limit to this power, in slowly and beautifully adapting each form to the most complex relations of life."

So then, if to a Naturalist, "the world is understood without recourse to spiritual or supernatural explanations," why then does Darwin personify nature here in a feminine gender by ascribing to it more than even the mere intelligent qualities and abilities of mankind? Moreover, why does his concept of nature portray the unlimited power and abilities of a god? For Darwin himself questions, "What limit can be put to this power," acknowledging here that it is this unlimited power of nature who is "rigidly scrutinising" and "favouring" the useful variations under changing conditions for "her living products," and who acts during long ages to selectively control the constitution, structure, and habits of each creature so that the whole process of evolution is achieved by "slowly and beautifully adapting each form" from the simple "to the most complex relations of life" while favoring the good and rejecting the bad. This may have been the origin of what has become commonly referred to as Mother Nature.

I strongly recommend reading Charles Darwin's autobiography, but please note that his family edited his autobiography. And if you read his autobiography, pay careful attention to his childhood antics, for Darwin describes himself doing many odd things just to get attention and to impress people. I found the autobiography to be very enlightening in helping to understand who Charles Darwin was, and what things influenced him along the way.

The perceptions of one man can sometimes have a powerful effect on many people, as demonstrated by the *Origin of Species.* Its influence has caused many to turn away from the Biblical worldview. The Bible teaches that we are to be responsible listeners. Take heed therefore how you hear.

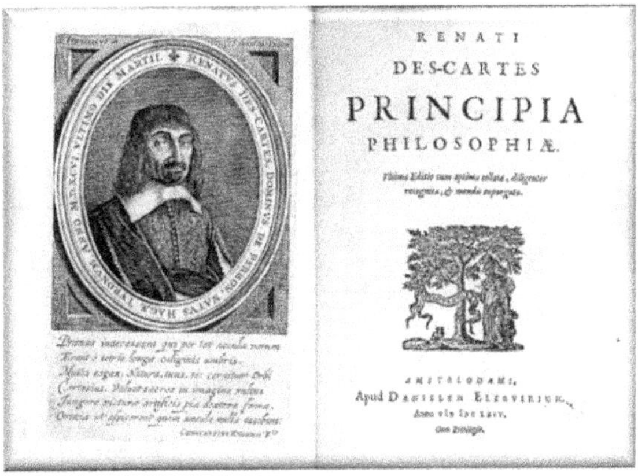

Chapter Four
Philosophy's Foundation

Some people think Modern Day Western philosophy originated with René Descartes. They make this assumption likely because he is so often referred to as the Father of Modern Philosophy. But those who have studied Early Western philosophy might suggest that the Early Western philosophers such as Socrates, Plato and Aristotle birthed Modern Day Western philosophy. They could say that since René Descartes, who came out of the Renaissance, a time when Early Western philosophy was actually rediscovered and had skyrocketed in popularity on a more global scale in European culture, used deductive reasoning, a technique that had become very popular during the peak of Greek culture, and which technique is plainly exhibited in all of the philosophical works of Plato[10] and more subtlety in the scientific works of Aristotle[11] whom he also admired.

It appears that Early Western philosophy may indeed be the heart and blood vehicle that continues to drive Atheism, Naturalism and Humanism today in the US.

Philosophy ▶ **Atheism** ▶ **Naturalism** ▶ **Humanism**

However, although it may be the vehicle, Early Western philosophy is still not the origin of modern day philosophy. If we rely on historical writings to determine the source, then the origin would likely go back to the Biblical account of the fall of man in the Garden of Eden,[12] and even beyond that, to the Biblical account of the fall of Satan from heaven,[13] since chronologically the fall of Satan proceeded the fall of man.

The technique used by Satan in the Garden with Adam and Eve was the subtle power of questioning the belief of

---

[10] http://classics.mit.edu/Plato/ [see the hyperlink to the translated text of each work]

[11] http://classics.mit.edu/Aristotle/ [see the hyperlink to the translated text of each work]

[12] Genesis 3:1

[13] Isaiah 14:3-15 [Satan, referred to here as the king of Babylon, oppressor, morning star, and the son of the dawn]

another in order to dismantle or disqualify it. Sound familiar? Socrates used this similar technique, which can be observed in many of the written works of Plato; but Satan was indeed the author of it, or for those who believe the story of Adam and Eve was merely a myth, Moses then was the author. Satan was also the originator of Naturalism and Humanism as can be observed in the Bible's accounts of the fall of man and the fall of Satan respectively, but since Satan wasn't human, it would probably be more accurate to call Humanism *Selfism* instead, or self-worship. Remember his five "I will" statements?[14]

Due to the renewed popularity of Early Western philosophy during the Renaissance, Early Western philosophy became integrated into the developing sciences, and it remains so to this day, where human intellect and reason, biased by these atheistic doctrines, have now become the prime measuring stick of interpretation within the framework of science.

Theological or God-given ethical directives were replaced by human reason based on Naturalism. But politicized religion had for centuries been perverting those God-given ethical directives anyway, so together, the naturalists and theologians have *both* served to taint truth and

[14] Isaiah 14:12-15

ethics for a very long time, and because Theology appears to have failed, it's no small wonder that Naturalism emerged within the early Greek culture and during the Enlightenment of the 18th century, to esteem the natural intellect and goodness of man as superior to the theology of Grecian gods and the Church.

History has shown that churches have upheld ignorant teachings that were not based on truth. For example, Descartes believed that the Catholic Church was making false assumptions and judgments about science and that the church was demonizing many scientists of his day; so he, as a Catholic, attempted through mathematics and philosophy to influence and affect the way the church thought about science by promoting a new philosophy that utilized deductive reasoning based on what Descartes[15] believed to be the most fundamental truth no one would disagree on; for who could deny, "I think, therefore I am?"

It became the common denominator and the foundation on which to build his new philosophy in order to change religious opinions about science. And it did much more than simply change religious opinions about science; it ultimately shook the confidence and beliefs of so many of those who had

---

[15] *The Philosophical Writings of Descartes*, translated by Cottingham, Stoothoff, Murdoch and Kenny, Cambridge UP, 1991.

once relied solely on the teachings of their religious leaders, and thereby served to convert them from believers in God to believers of science.

It was good that the science of that time served to expose the corruption that existed within the religious political structures. However, the net result was that since the beliefs of many were shaken by the enlightenment of science, science then became the replacement for religion, something more tangible to believe in, and something that certainly didn't appear on the surface to require any faith.

Chapter Five
Realization of Truth

René Descartes believed that no rational person would argue his own existence (he obviously had no idea what roads philosophy would take in the twentieth century and beyond). In fact, Descartes described himself as having ridded himself of all doubt in order to arrive at the indisputable fact of his own existence. That premise became the substance for his common-ground agreement with all rational people. All doubt gone—indisputable fact remains—truth.

René Descartes highly valued reasoning. Logic was the means he used to convey his new philosophy. Truth therefore had to be his premise, because if "I think" was not a true statement, then "Therefore I am" would be untrue as well. No truth – No René Descartes, or at least no premise for him to build upon. It's as simple as that. He may not have ever stated it so plainly, but it certainly was an assumption that was implicit in his writings.

But consider this—that after Descartes eliminated all doubt, only truth remained. It wasn't something he added; truth was already there—the fact of his existence. He merely removed all the doubt dirt covering it up, and in doing so uncovered the diamond that was hidden beneath it. Doubt gone—truth remained. Descartes *indeed* existed, he was certain of it.

But think about this: If it was true that René Descartes existed *when* he realized it, then it was certainly true that he existed *before* he realized it, for he had lived many years as a boy before realizing the truth of his own existence, and before recording it on paper. So if it is true that Descartes existed before he realized it, then truth also must have existed prior to Descartes' realization and perception of it as well.

And, if this were true for René Descartes, then this would also hold true for everyone else also, including the very first human being on planet Earth. So, since we can assume that the first human being on earth had to have existed prior to his ability or opportunity to perceive and acknowledge his own existence, then it is equally rational to assume that truth must therefore exist apart from man's perception and acknowledgement of it; And with this concept, we have a sound premise to build from:

TRUTH EXISTS APART FROM MAN'S PERCEPTION AND ACKNOWLEDGEMENT OF IT

Chapter Six
Evidence of Truth's Existence

Our perceptions are very limited at best, which is why it is so important that truth be the fundamental goal of the individual, and the prime objective for the advancement of science. Truth supports everything that exists within the universe. It affects all organic, inorganic and immaterial things as well: rocks, trees, plants, animals, people, concepts, stories, places and events in ways that will be discussed.

We have already established that truth exists apart from man's perception of it. Next it is important to understand that truth *is* what it is. It stands alone. It is not what *we* believe it to be unless our beliefs are aligned with the truth. This concept leaves our perception of the truth always under scrutiny, and not the truth itself. Therefore, the pursuit of truth should become the unquenchable quest and the foremost focus for every individual, especially for believer scientists and students of science.

Logic is an excellent tool to use in the pursuit of truth, or for the understanding of truth. But it is equally important to recognize that *truth may exist without logic, but logic cannot exist without truth, or at least the assumption of truth.* In order for logic to exist, there must be true and false, yes and no. Computers and software applications could not operate without the binary ones and zeros. They rely on it – true and false – if and then.

Have you ever wondered what it would be like if truth didn't exist? Imagine it for just a moment: You would not be able to depend on anything because consistency would not exist. You would be very afraid because you couldn't trust anything – one moment something might be safe, and the next moment it might be harmful. Then, since you wouldn't be consistent either, you might start to not be afraid anymore and trust something you shouldn't. Bees might sting you one minute, and chirp like a canary the next. You couldn't depend on physical laws such as gravity, motion, electricity, or chemical bonding because they wouldn't exist either. You wouldn't recognize anything because everything, including yourself, would be constantly changing. In fact, a scenario about life without truth sounds so ridiculous that I'm embarrassed to even write about it. If you think you can write one that sounds believable, I'd appreciate receiving a copy.

Thank God for physical laws! Also, you may want to consider this: You wouldn't exist either, if your body, soul and spirit didn't hold together. The very fact that substance holds together is the *only* reason there is physical order in the universe. How is it that you can construct thought patterns in your brain in the form of intangible concepts, ideas and imagery, and then store them for future use? Truth and logic require consistency and order, so without order there is no truth or logic.

THE FACT THAT ALL THINGS HOLD TOGETHER AND ARE ORDERLY IS EVIDENCE OF TRUTH'S EXISTENCE

When I was much younger, I used to catch myself thinking in philosophical loops as I pondered my own existence: How do I know I really exist? I know I'm aware that I exist, but how can I know this is not an illusion? How can there be a God that always exists, who had no beginning and has no end? If he does exist, how can I know he is not just playing games with his creation? How can anything exist at all? Yet, how can nothing exist? What is something, and what is nothing? Why am I here? Why wasn't I born in Africa or

China? Who am I? What am I? Is mankind simply atoms that came together from primordial soup that formed by happenstance, with no recipe and no chef to measure and mix the ingredients? What is life? What is death? Is there life after death? Do I exist forever?

Too much of this type of thinking can make you crazy, or at the very least it can be very disturbing. It is best to start with the basic concept of truth itself and take it from there—it is more stabilizing anyway.

When a person realizes that truth actually exists independently of what people think, and that truth is what it is—whether it can be explained or not, we find stability in this knowledge of truth's existence, and we can then build logically and confidently on that knowledge to become anchored in that truth. "…I like to think that the moon is there even if I am not looking at it." —Albert Einstein

These days, everyone seems to have his or her own opinions concerning what truth is and how it can be found, and it tends to be defined most often according to whatever a person perceives it to be. Below are a few common statements one often hears concerning truth:

- Truth is relative.
- It's only truth if you believe it.

- What truth is for one person may not be truth for someone else; and that's okay, it's *okay* to disagree. People see things differently.

- The truth is somewhere in the middle.

- Truth is hard to find, but you'll recognize it when you see it.

- The truth is what makes the most sense.

However, if truth exists apart from man's perception and acknowledgement of it, then surely it *is* something – but *what* is it really? And how can we know what it is?

Even in the US Justice System, where it is all or nothing on behalf of the jurors in order to protect the innocent, we still wonder: were the jurors *really* correct in their verdict? Did they actually hear "the truth and nothing but the truth" from the witnesses? Were the facts untainted that were presented to them? Was there reasonable doubt? What *is* reasonable doubt anyway? Was everyone's concept of a *normal, prudent person* the same?

I think an unspoken question people often silently ask themselves is, "How can I be sure this information is really true?" Although many truths are self-evident or even researchable, many are not. But truth-detecting skills can be learned. Judges often become highly skilled in recognizing when people are telling the truth and when they are not. Body language plays a major role in revealing the truth of someone's conversation. Concerning truth: if one person is right, does the other person always have to be wrong? It certainly would be comforting to many if everything were always black and white, right or wrong, true or false—but truth is not always that way.

Truth doesn't always mean that only one side of an argument or statement is true. Many times opposing statements are both true. For example, I can say that I'm *sad* my father died. But I can also say I'm *glad* my father died, and both statements are true even though they sound so opposite. Haven't you ever been both glad and sad at the same time? I'm sad that my father passed on because I am no longer able to see him, to talk with him or to hug him or even hold his hand. But I'm also happy that he passed on because he is no longer suffering, and because I know he is in a better place right now. This point will be clearer as we look at what truth actually is in a more precise manner.

Chapter Seven
What Is Truth?

According to René Descartes, there was one indisputable fact for a person in his right mind: once all doubt had been removed, the indisputable fact of one's own existence remained – truth remained.

Because truth *is* what it is, and it exists, it should therefore be perceptible and recognizable. Unfortunately, truth is generally invisible, and misperceptions and false doctrines often make truth difficult to discern as well, just as fool's gold can be a distraction to finding genuine gold. So how can we more easily distinguish the truth apart from all the lies, false perceptions, and the fact that it so often seems intangible?

Truth has certain characteristics that make it distinguishable. It is very similar to gold in that you must dig deeper to find it. But when you make the effort to clear the obstacles that so often conceal truth, it suddenly appears like specs of gold shimmering in a gold pan. As gold becomes more pure being subjected to bellowed fire, so truth withstands a broadside of intense accusation and questioning.

I once learned that gold is often found with quartz, and when there is also iron present in the soil; and when there are also traces of silver or platinum, there is a very good chance there may be gold as well. These are some of the signs generally present when and where gold has been found. Truth is found most often where there is consistency; and where there is repetition or patterns present, truth is generally close by. People who live according to the truth are generally recognized by their consistency, whereas those who do not live according to the truth are usually recognized by their inconsistent, problematic lives often accompanied by many excuses.

It's not easy to live a lie. When someone chooses to live a lie or cheat, they must struggle to cover all the bases so that their lies or cheating won't be revealed, which generally involves telling more lies. The more the lying continues, the more stress is experienced by that individual, and the more stress in a person's life often leads to more illnesses.

Those who choose to lie often also live a life of fear—fear that they'll be caught or revealed. They work so hard to deceive in order to conceal the truth. And yet, a little slip of the tongue, a little bragging about how they got something by their craftiness, and they suddenly find themselves in trouble because of honest people who hear it and report them.

Ephesians 4:15 says we are to speak the truth in love. It often opens you up to potential judgment by others if you are too truthful, or you can easily hurt others by openly sharing the truth you know if you are indiscrete. You must be careful how you hold or use the truth, and always make sure that you are operating in love. This means you are responsible for what you do with the truth. Truth certainly makes its own enemies, and the Bible cautions us to be careful *not* to cast our pearls before the swine, for if we do, they may get trampled on.

It certainly feels good on the inside to live a truthful life, for it brings freedom and peace within. The Bible tells us to put on the belt of truth. Always speaking the truth is our protection, so we don't get caught with our pants down, so to speak, in a lie. It keeps us unashamed.

Dictionary Definition

But what actually *is* truth? The Merriam-Webster Online Dictionary[16] defines truth as:

1 a *archaic :* FIDELITY, CONSTANCY
b : sincerity in action, character, and utterance

2 a (1) : the state of being the case : fact
(2) : the body of real things, events, and facts : ACTUALITY

---

[16] Merriam-Webster Online (http://www.merriam-webster.com/dictionary/truth); Merriam-Webster Online Dictionary copyright © 2011 by Merriam-Webster, Incorporated.

(3) : *often capitalized* : a transcendent fundamental or spiritual reality

b : a judgment, proposition, or IDEA that is true or accepted as true <*truths* of thermodynamics>

c : the body of true statements and propositions

3 a : the property (as of a statement) of being in accord with fact or reality

b *chiefly British :* TRUE 2

c : fidelity to an original or to a standard

4 *capitalized Christian Science* : GOD

— in truth
: in accordance with fact : ACTUALLY

The definition seems to be very representative of the actual modern-day usage of the term *truth.* However, part 1b indicates that the "Sincerity in action, character, and utterance" is considered to be truth. One could make argument that there are many people who are very *sincere* about the things they believe, what they say to others, and about the way they behave regarding certain things. They may have truly said or done something with all sincerity, but that doesn't necessarily mean that what they said or did was actually true or correct.

In part 1a, truth has been described as "fidelity" and "constancy." Constancy is an excellent concept; it means, "A state of being constant or unchanging." This is a good concept to embrace.

Part 2a(2) of the definition makes the statement, "the body of real things, events, and facts: Actuality." This is an excellent fragment describing truth. It identifies truth as not being false or apparent, but existing in fact, in reality, and not merely potentially, and therefore, truth *is* what it is.

Its reference to the *body of real things, events, and facts* is also an important concept, because truth, like a *body* having many parts, often consists of many sub-events or sub-facts that together form the complete body of truth regarding a particular subject or event. These sub-events or sub-facts can be viewed separately as perspectives, that when considered all together they make up the entire truth about a particular subject or event.

Part 4 of the definition references the word GOD, as when it is capitalized in *Christian Science*, to mean truth. Many people, both religious and nonreligious, would probably object these days to this part of the definition having been included; but it is indeed interesting when both God and truth are found to be synonymous, especially within a popular dictionary.

So, we have a dictionary definition of truth that includes descriptive words such as "Fidelity," "Constancy," "Actuality" and "GOD." It would probably be good to point out here that except for the constancy concept of truth, a *time* element or concept of truth is not even included in any of these dictionary examples of truth. But a time concept of truth is very important, because some truths are related and connected with time or timing. The next chapter is a unique way of looking at truth. It is a method of breaking down truth in order to better understand it, so these important time elements are included.

As you read the next chapter, please keep in mind that truth *is* what it is. We can only view it from certain perspectives in order to attempt to understand it. But man is very limited at best, and therefore, it is always good to keep it in mind as well.

Chapter Eight
Aspects of Truth

As you consider truth, it is important to realize that truth can often be broken down into many fragments, or perspectives. A *Perspective truth* is a truth fragment from a particular vantage point or layer of consideration. Truth from one perspective may appear to be in conflict with truth from another perspective as will be shown in the White Ball example to follow, but Perspective truth is *always* accurate from the particular perspective or else it is not truth at all. Perspective truths will harmonize and make sense in light of the bigger picture, if they are indeed true and accurate perspectives.

Perspective Truth I
White Ball Example

Imagine you are sitting on a chair near the center of a cube-shaped room with nothing but a small, white ball in front of you that appears to be moving in a vertical, circular motion in the center of the room as if it were attached to the outer end of the second hand of a wall clock. It is cycling around and around in a clockwise direction, making one revolution per second and leaving a short vapor trail that quickly dissipates, but lasts long enough so that you, the observer, can see where the ball has been.

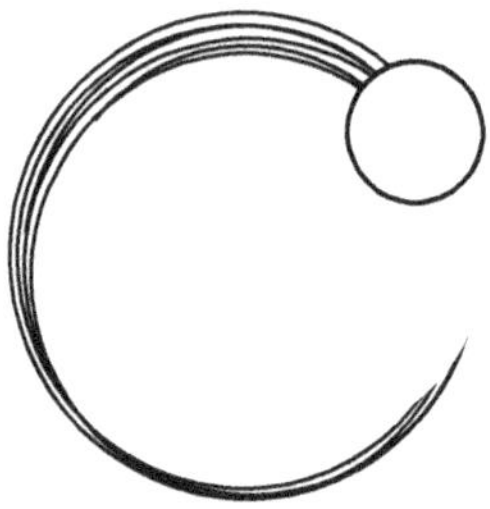

If you are asked to describe the basic motion of the white ball, since a white ball and its vapor trail are all you see, you might describe it as moving in a circular, clockwise direction. To someone else seated directly opposite from you, on the other side of, and behind the same white ball, they might describe it as moving in a circular, counter-clockwise direction. But just because you say it is moving one direction, and the other person says it is moving the opposite direction, it does not mean that one person is telling the truth, and the other is not. Both are telling the truth, according to each person's perspective.

Now had the person behind the white ball described the motion of the ball to also be in a clockwise direction, then that person would *not* be telling the truth, and therefore, the false statement would *not* be considered Perspective Truth at all. In order for *perspective truth* to be such, it must be perceived and represented accurately from the particular perspective. Now let's get back to the white ball.

To someone seated to the right or to the left side of the white ball relative to you, they might describe the white ball as moving up and down in a vertical direction. To someone seated above or below the white ball and looking up or down at it respectively, they might describe the white ball as moving from side to side in a horizontal direction.

Which of these six different perspectives most accurately describes the truth about the motion of the white ball? If you were thinking all of them do, you would probably be correct, except for the fact that there is something you do not know because you were not told. The cube-shaped room is on the flat car of a train traveling from South City to North Town. But being inside the cube, you didn't know. There are no passengers at the Midway Train Station, and since the train is only pulling a flat car, the train will be passing by the station without stopping.

If the cargo on the train were invisible except for the white ball and its vapor trail, and someone inside the train station were to glance out the window at that very moment when the white ball on the train was passing by, that individual might describe the motion of the white ball as appearing somewhat like a sine wave, based on its vapor trail.

If a worker making roofing repairs and working high up on the train station roof at the time when the train was passing by were to turn around and look at the train just after it had passed the station, that worker might describe the motion of the white ball as moving like a coil, based on its vapor trail.

Now which of *these* two observations most accurately describes the motion of the white ball? Do you think that both are true according to their perspectives? Great! However, because those inside and on top of the train station were unaware of the effects that the rotation of the earth, the wobble of the earth, and the orbit of the earth around the sun all play on the actual motion of the white ball, then each observation is true, but neither perspective *accurately* describes the absolute motion of the white ball.

And what about the effect that our solar system has as it moves within the Milky Way galaxy? What effect does it have on the motion of the white ball? Also affecting the motion of the white ball would be the movement of the Milky Way galaxy within the larger universe? And God only knows what else may be affecting the motion of the white ball!

Perspective Truth II

Wild Bear Example

People and animals function similarly. A wild bear, for example, has an instinctive fear of man. The bear naturally believes man is a threat. But in a national park setting where the animal life is respected and protected, the bear's own perception and experience within its environment can alter this naturally protective fear instinct.

As a wild bear may be subjected to many encounters with humans in this protected environment, it adapts as it believes man is not a threat after all. It soon learns it can bash car windows and upset campsites to get its treats, and since man poses no real danger, the bear often becomes bolder and more dangerous to those visiting the park. The bear doesn't understand that in the park, bears are protected by law. The laws change the people's behavior, and so the bear gets his picnic basket. Except that…eventually, the unruly bear finds itself captured in a live trap by rangers, and transported back to the wilderness, a more difficult living environment.

So then, what the bear *thought* was true, was only based on its own limited perceptions and experiences, and not the fact that the National Park laws were governing human behavior. The bear did not understand the bigger picture. After having enjoyed the delightful tastes of human foods for a season, it must now suffer the adjustment of reverting back to the bland, natural foods diet that requires so much more effort to obtain.

We should learn from the wild bear that although our perspective may be true, that perspective may not be all there is to the matter. It is always important to keep an open mind, and to keep looking for the *bigger picture*, realizing that the bigger picture is generally achieved by collecting data from as many perspectives as possible.

The Bigger Picture
The Importance Factor

From the examples of the White Ball and the Wild Bear, we can see that any perspective may be true from that perspective, but does this mean all truth is relative? No. Truth *is* what it is – the white ball, if it would have been real, would have had a true motion that could have only been most accurately described by the truth from all perspectives. Does this mean truth from only one perspective is a lie? No. It is still true for that perspective, and may be useful within that perspective and beyond.

The perspectives within the cube are useful to those inside the cube. Perspectives from the earth are useful to those living on the earth. Does this mean truth from more than one perspective is better? Yes. The more truth we obtain from a variety of perspectives, the more likely we are to see the more accurate bigger picture of what is being observed.

Perspective truth may be beneficial to the observer, depending on the situation. If I only look at my rear view mirror perspective while driving a car, it could cost me dearly, and possibly harm others. There seems to be an Importance Factor associated with the issues of life: The more important things are, the more necessary it is to gain as many true perspectives as possible. It's just common sense really.

The Human Factor

False Perceptions

Perspective Truth is only truth if it is correctly and accurately perceived. As humans, however, we do not always see what is true from our perspectives. Our perceptions may be tainted or exaggerated by our biases and hopes. We may be limited by varying degrees of sight, hearing, knowledge, and understanding, or by misinformation or hearsay from others.

What happens after we die? How many perspectives should a person have of this? What is the Importance Factor here? Remember, hell either *is*, or it is not. I can believe that hell exists, or I can believe otherwise. My belief that Hell doesn't exist won't change the fact that it does, for truth *is* what it is. What a person believes about truth will never change the truth—but truth can certainly change a person.

If I believe Hell is only what others have said about it, I may be in for a shock. Or if my perception consists of what I have made up in my mind because I believe God is only a loving and kind God, then there should be no concern even if I believe in Hell, as long as I do whatever I believe a loving, kind God would want me to do in order to avoid it. But that could be a grave mistake if I fail to consider what the one who created Hell actually has to say about it.

It would be a mistake to ignore the parts of the Bible that pertain to God being a just, holy, and righteous God, one who punishes evil, and says, "Depart from me you cursed, into the lake of fire prepared for the devil and his angels." When Jesus says, "I am the way, and the truth, and the life. No man comes to the Father *except* through me,"[17] it would be most wise to consider *who* it is that is actually saying that, and to consider everything he has to say about it.

In other words, what we perceive may not even be true at all, and if not, then it does not even fall under the category of Perspective truth. History books are filled with perspectives, but unfortunately, most are misperceptions and conjecture built on hearsay or exaggerated facts. But whatever historical events are printed that accurately describe events that have occurred, these would be considered Perspective Truths, and since they occurred at or during a specific time period, they would also fall into the classification of Time-related truth as well.

The point here is that it is essential that we don't become so self-important that we forget the *Human Factor*—

---

[17] John 14:6 NIV

that is, just how vulnerable we really are as human beings, and how susceptible we are to making mistakes, drawing irrational conclusions, and to developing many misperceptions built on hearsay.

We need to highly value truth, and to seek God's help in discerning whatever is true. Remember, *false perceptions lead to false conclusions*. When someone perceives something differently because of a new perception, it often manifests itself in the form of changed behavior or actions. New or better information often leads to better understanding and improvement. False information can lead to erratic behavior, and ultimately end up in disaster.

Also consider this most important fact—the word of God trumps what we perceive to be our reality. If your situation appears to be one way, but the word of God says something different about it, which reality will you believe? For example, when finances have me down, and when I cannot see how I am going to pay for all the things I am responsible for, I have a choice. I can either believe I can't do it, or I can believe that with God all things are possible.[18] I can believe that my God shall supply all of my needs according to his

---

[18] Matthew 19:26; Mark 10:27

riches in glory in Christ Jesus,[19] and I can believe that my God shall do exceedingly abundantly above all that I ask or imagine.[20] I can choose to live by faith, and not fear.

Time changes a pollywog into a frog, and it changes a caterpillar into a butterfly. These creatures must simply eat and survive in order to realize this amazing metamorphosis and to become what God designed them to be. When we believe what God says, and press on, believing that he is able to accomplish what he promises, then we also can experience an amazing metamorphosis in our own perceived realities as well.

The human factor is not an issue where faith in God's word is involved. My wife and I have personally experienced metamorphic transitions in our lives financially and otherwise as we've looked to God for the bigger picture.

Time-related Truth

Time-related truth is truth, or Perspective truth, that is associated with some element of time or timing. Time-related truth may be a repetitive or cyclical pattern that keeps recurring at measured intervals, or that lasts for a specific period of time, and then ceases. Time-related truth was, is, or will be true for a

---

[19] Philippians 4:19
[20] Ephesians 3:20

specific time, period or parameter of time, but afterwards will never actively be true again.

Time-related truth may also be considered from an eternal or time-independent perspective in that once it has occurred and afterward subsides, it then is historically true forevermore, and its effects may either be everlasting into eternity, or else insignificant and oblivious thereafter.

Time-related truth is like something wrapped in a package. It can be many things. It can be a sequence of events that happened in the past—a crime for example. Detectives spend lots of time carefully gathering evidence connected to the events that occurred during a specific time period. They search for truths from as many perspectives as possible in order to get the bigger picture of what happened, and to produce a preponderance of evidence in order to obtain a conviction.

Time-related truth can be a senior citizen discount that is only applied when a person reaches a specific age, but not until. It can be a limited time opportunity, a sale that ends on Sunday, or a job offer that opens on a particular date, and closes on the fifteenth of the month. It can be a series of anniversaries, each occurring once per year until it is no longer celebrated, or it can be an individual's heartbeat that beats rhythmically for a lifetime, and then one day is silent forever.

Eternal Truth

Time-independent

Eternal truth is truth, or Perspective truth, that remains the same regardless of time. It was, is, or always will be true, or else it never was, never is, or never will be true. Eternal truth may be a repetitive or cyclical pattern that recurs without end, or it may be a Time-related truth observed and considered from an eternal perspective where time is no longer relevant.

Eternal truth is changeless, or it is always changing. Energy, for example, is Eternal truth—it always exists, it is never lost, but it forever changes form, location, concentration, and manifestation.[21] Space is also Eternal truth, infinitely occupying all regions of the cosmos, without beginning and without end; containing all things that exist, including all objects, planets, stars and galaxies. Space is timeless and everlasting.

Eternal truth is God, for he is the same yesterday, and today, and forever.[22] He said, "Before Abraham was, I am."[23] He said to Moses, "I am that I am."[24]

---

[21] Law of Conservation of Energy, the First Law of Thermodynamics: see http://en.wikipedia.org/wiki/Conservation_of_energy
[22] Hebrews 13:8
[23] John 8:58
[24] Exodus 3:14 (KJV)

Eternal truth always was, is, and always will be, or else it never was, is not, and never will be. Eternal truth is like the circle of a ring that has no beginning and no end. It is unaffected by time. It exists in an eternal now, a present tense without end. It is timeless.

Absolute Truth

Absolute truth is the sum total of all the Perspective truths surrounding a particular subject or event, collectively making up the entire truth about any particular subject or event, just as words may be logically arranged to form a meaningful sentence. A logical sentence is made up of many words, each word having individual purposes and meanings that collectively contribute to the overall meaning of the complete sentence.

God, who is truth, and who is omniscient (all-knowing) and omnipresent (all-present)—He is the eternal witness who observes all things from all perspectives. He alone is able to testify regarding the absolute truth about anything and everything, for he alone is unconstrained, functioning inside and outside of time. He is both timeless and timely. He is the great I AM. This is why it is so important that we become familiar with God's written word because it is an abundant resource of truth from the One who knows truth absolutely.

Man cannot be in every place at the same time like God is. Man does not have the all-seeing eyes of God. Man cannot change the fact that something once happened, or that something once was. Man can only change the now to affect the future, or to affect the history that will be written afterward. So, as long as man is unable to observe all things and events from every perspective, he may never know the absolute truth about anything in this world or the universe as perceived through his own senses.

Therefore, it is imperative that God be both our teacher and our guide by asking him for guidance and direction in all studies, investigations and explorations, especially in a world filled with so many wonderful things that he has created for us to know and to use for our health, pleasure and enjoyment.

So then, Absolute truth is the complete truth regarding any particular thing or event, incorporating Time-related and/or Eternal truth where applicable.

Chapter Nine
Truth and God

So, after having ridded himself of all doubt, René Descartes reasoned that because he was thinking, he must therefore also exist. But since Mr. Descartes was a man when he realized it, he must have existed before he realized it, for he was once a child. So the truth of his existence preceded his realization of it, and therefore truth must have existed before he realized it as well.

If that was true for René Descartes, then it must also hold true for mankind, and so truth also existed before the very first man could have ever perceived and acknowledged his own existence. Therefore, truth existed prior to mankind's perception of truth. From this we deduced that *truth exists apart from man's perception and acknowledgement of it.*

Therefore, it is logical to conclude that *truth is what it is*, and that truth is independent. Although it may seem farfetched to some, it would not at all be illogical to also suppose that if truth *could* speak for itself, and that if truth were asked to describe itself from its own perspective, it would have to say, "I am that I am." Does that sound familiar? When

Moses didn't know what to tell Pharaoh, and he asked God the question, "Whom shall I say sent me?" God answered Moses, "I am that I am..."[25]

Secular logic, then, would not be in disagreement with the Bible's declaration that God *is* truth. A human being might argue concerning himself or herself, saying, "I am that I am too," but because every human being is inconsistent, it would be a lie, and therefore, the only truth about that statement would be the fact that human beings are inconsistent. Only God and truth are absolutely alike. They are one and the same.

Let's take it one step further: What about space? We know that space exists, because objects occupy space. There are objects occupying space all over our planet and to the furthest reaches of our galaxy and beyond, throughout the entire universe. But there is also inner space, consisting of those micro spaces between atoms and quarks as well. The existence of space is real, even according to modern science, so therefore, the fact that space exists is truth, and since space is everywhere, then truth is also everywhere. Truth is therefore omnipresent. This is also another attribute of God according to the Bible.

---

[25] Exodus 3:14

So, we have considered Perspective truths as being fragments of truths or events that can be time-related, eternal, or both. Perspective truths are never false, but are always accurate or else they are not truth at all. Time-related truth can also be Eternal truth when viewed from an eternal perspective, or if its effects are everlasting after its manifestation.

God is an eternal being who is omniscient (all-knowing) and omnipresent (all-present), and therefore he is the only true witness to the absolute truth about anything and everything—for God, and God alone, observes all things from all perspectives.

GOD ALONE IS THE ETERNAL, TRUE WITNESS,
WHO OBSERVES ALL THINGS FROM ALL PERSPECTIVES

Part II: Truth from Biblical Perspectives

Chapter Ten
What the Bible Says About Truth

God's written word is the most valuable resource for truth. If you know what the Bible has to say about various subjects, you will find that many questions have already been answered. For example, the age old question, "What came first, the chicken or the egg?" It's simple— the chicken! Because God created "…every winged bird according to its kind," and *then* God said, "…let the birds increase on the earth" (Genesis 1:21-22).

There is nothing in the Bible to indicate that God first created bird eggs. Besides, we now know that the DNA within the fertilized egg contains all the genetic programming for a hatchling to become a complete adult chicken and utilize its God-given ability to share in the reproduction process of fertilizing or producing and laying eggs. Besides—how can anything be reproduced unless it is first produced?

The Bible has a great deal to say regarding truth. Here are just a few categorized things the Bible says about truth (All passages are taken from the NIV Bible):

God *is* Truth, and He is the God *of* Truth

John 14:6
Jesus answered, "I am the way and the truth and the life. No one comes to the Father except through me.

Psalm 31:5
Into your hands I commit my spirit; redeem me, O Lord, the God of truth.

John 14:16-17
And I will ask the Father, and he will give you another Counselor to be with you forever— the Spirit of truth. The world cannot accept him, because it neither sees him nor knows him. But you know him, for he lives with you and will be in you.

1 John 5:6
This is the one who came by water and blood—Jesus Christ. He did not come by water only, but by water and blood. And it is the Spirit who testifies, because the Spirit is the truth.

Truth Liberates

John 8:32
Then you will know the truth, and the truth will set you free."

Truth is Accepted or Rejected

Romans 1:25
They exchanged the truth of God for a lie, and worshiped and served created things rather than the Creator—who is forever praised. Amen.

Romans 2:8
But for those who are self-seeking and who reject the truth and follow evil, there will be wrath and anger.

Truth Belongs to God

Psalm 40:10
I do not hide your righteousness in my heart; I speak of your faithfulness and salvation. I do not conceal your love and your truth from the great assembly.

Psalm 40:11
Do not withhold your mercy from me, O LORD; may your love and your truth always protect me.

Psalm 25:5
Guide me in your truth and teach me, for you are God my Savior, and my hope is in you all day long.

Psalm 26:3
for your love is ever before me, and I walk continually in your truth.

Psalm 96:13
They will sing before the LORD, for he comes, he comes to

judge the earth. He will judge the world in righteousness and the peoples in his truth.

The Word of the Lord is Truth

1 Kings 17:24
Then the woman said to Elijah, "Now I know that you are a man of God and that the word of the LORD from your mouth is the truth."

Truth Brings Man Closer to God

John 3:21
But whoever lives by the truth comes into the light, so that it may be seen plainly that what he has done has been done through God.

John 4:23
Yet a time is coming and has now come when the true worshipers will worship the Father in spirit and truth, for they are the kind of worshipers the Father seeks.

John 4:24
God is spirit, and his worshipers must worship in spirit and in truth.

He Guides Us to Truth & Speaks Truth to Us

John 16:13
But when he, the Spirit of truth, comes, he will guide you into all truth. He will not speak on his own; he will speak only what he hears, and he will tell you what is yet to come.

John 18:37
"You are a king, then!" said Pilate. Jesus answered, "You are right in saying I am a king. In fact, for this reason I was born, and for this I came into the world, to testify to the truth. Everyone on the side of truth listens to me."

John 15:26
When the Counselor comes, whom I will send to you from the Father, the Spirit of truth who goes out from the Father, he will testify about me.

When you consider what the Bible has to say about truth, consider also the many stories found in the Bible that teach us about truth and the importance of being truthful. The four gospels alone give incredible examples of how important truth is to God, and they show us plainly how Jesus lived a life of truth despite all the adverse circumstances he experienced as the result.

The bigger picture begins with understanding truth itself—realizing that God and truth are inseparable, and by recognizing that God is the primary source of all truth and wisdom. The Bible is filled with truth. If we lack wisdom, we should go to God for it. For the Bible says, "If any man lacks wisdom, let him ask of God who gives to all men liberally and doesn't uphold."[26] God wants us to have wisdom and understanding. So what could be better than to consult with the source of all knowledge and wisdom—God himself? Remember, the Bible says, "You do not have, because you do not ask God."[27]

Truth of God's Creation

Begin With a Biblical Perspective

One thing you may want to consider as you conduct research or investigation in order to find the truth about any particular topic is this: *Begin with a Biblical perspective.* This means that you should study the scripture first hand to see for yourself what the Bible actually says, rather than simply researching what others have said it says.

First go to the source, and *afterward* compare what God has shown you with what others may have to contribute to the subject. As an example, if you want to know the truth about Charles Darwin, then you should read what Charles Darwin has written first, and not what others have written about him. It is good to form your own preliminary opinions before consulting

---

[26] Insert Bible quote: If any man lacks wisdom, let him ask of God who gives to all men liberally and upbraideth not.
[27] James 4:2 NIV

others for their opinions. Parrots are a dime a dozen, but independent thinkers are too few.

The Bible is very clear and forthright about many subjects, but some things are not so obvious. For example, it is very clearly stated in the Bible that God created the heavens and the earth,[28] and that he stretches out the heavens with his hands.[29] Those Biblical facts are repeated many times throughout the scriptures, so you can quickly discern God's perspective on that issue.

On the other hand, there are also many subjects that are not forthrightly obvious, and which can only be learned by carefully studying and knowing the Bible. By conducting a simple word use study to see how certain words have been used throughout the scriptures, a quick glimpse or perspective can be gained of the scope of meaning of particular words as they have been used throughout the Bible.

A search engine can be a very useful tool to use in obtaining the basic Biblical perspective of the earth's existence. It is common knowledge that the Bible says God created the earth. So by entering the word *create* in a search engine, and the word *earth*, you can quickly gain a broad

---

[28] Genesis 1:2 NIV; Isaiah 42:5 NIV; Isaiah 45:18 NIV; Revelation 10:6

[29] Job 9:8 NIV; Psalm 104:2; Isaiah 40:22 NIV; Isaiah 42:5 NIV; Zechariah 4:1 NIV

perspective of the Biblical account of when the earth was created. If you want an even broader perspective, you may also want to include, or separately search, the word *heaven* as well, to make sure you are getting the bigger picture as found throughout the old and new testaments.

Following are only a few scriptures that indicate God's perspective of how everything came into being. The Bible makes it very clear that God created everything that exists:

Genesis 1:1 [NIV used typically throughout this section]
In the beginning God created the heavens and the earth.

Genesis 1:21
So God created the great creatures of the sea and every living and moving thing with which the water teems, according to their kinds, and every winged bird according to its kind. And God saw that it was good.

Genesis 1:27
So God created man in his own image, in the image of God he created him; male and female he created them.

Genesis 2:4
This is the account of the heavens and the earth when they were created. When the Lord God made the earth and the heavens-

Psalm 8:3
When I consider your heavens, the work of your fingers, the moon and the stars, which you have set in place,

Psalm 19:1
The heavens declare the glory of God; the skies proclaim the work of his hands.

Isaiah 42:5
This is what God the Lord says— he who created the heavens and stretched them out, who spread out the earth and all that comes out of it, who gives breath to its people, and life to those who walk on it:

Isaiah 44:24
This is what the Lord says— your Redeemer, who formed you in the womb: I am the Lord, who has made all things, who alone stretched out the heavens, who spread out the earth by myself,

Isaiah 45:12
It is I who made the earth and created mankind upon it. My own hands stretched out the heavens; I marshaled their starry hosts.

Isaiah 45:18
For this is what the LORD says— he who created the heavens, he is God; he who fashioned and made the earth, he founded it; he did not create it to be empty, but formed it to be inhabited— he says: "I am the LORD, and there is no other.

Jeremiah 10:12
But God made the earth by his power; he founded the world by his wisdom and stretched out the heavens by his understanding.

Zechariah 12:1
This is the word of the Lord concerning Israel. The Lord, who

stretches out the heavens, who lays the foundation of the earth, and who forms the spirit of man within him,

Hebrews 1:10
He also says, "In the beginning, O Lord, you laid the foundations of the earth, and the heavens are the work of your hands.

Chapter Eleven
Relevance of Truth

Just because a person believes truth exists, it does not mean that the person knows the truth, or even knows what truth is. The Bible says, "You believe that there is one God. Good! Even the demons believe that—and shudder."[30] So, whether a person believes there is a God or not, according to the scripture we just read, belief in God only places them on an even keel with the demons.

And just because a person believes in Jesus Christ, it does not mean that the individual is a follower. If you know for a fact that jumping in the nearby swimming pool can save you if your clothing catches on fire, but you don't act on that knowledge when your clothing actually catches on fire, then you're going to burn. That is also true. The point is this, that if you know the truth, you would do best by acting on that knowledge, otherwise it is of little value to you, and ultimately it will not benefit you in the end.

---

[30] James 2:19 NIV

On the other hand, many people believe things to be true, when, in fact, they are not at all true. What we believe, and what the truth really is, can often be very different. Our beliefs are forever changing, but truth is not something that is editable. We may think we can rewrite it, like what we can do with words written on paper, but we cannot, because truth *is* what it is. We may be able to convince others, and even ourselves, that something is not the truth, when in fact it is, but that still doesn't change the truth, it merely hides or disguises it.

Perceptions are also varied depending on personality types. When asked to describe a glass containing liquid at the halfway level, the more optimistic person will generally convey a positive description by stating, "The glass is half full." The more pessimistic individual will most certainly convey the bleaker perspective by saying, "The glass is half empty." Most of us have heard that one before. But even though these are honest attempts at answering the question, only one of them can actually be correct. For in order to accurately answer the question without simply guessing it correctly, one must have either first observed how the liquid in the glass got to its present state, or else one must first ask the question, "What was the state of the glass before the liquid reached the current level?" If the glass was full to begin with, then the correct answer would be, "The glass is half empty." If the glass was empty to start with, the correct answer would be, "The glass is half full."

But when questions are ones of utmost importance, being an optimist is not always going to ensure you have an accurate answer. It is probably best to be a realist with regard to issues where truth is extremely significant, and to not allow your optimism to influence the accuracy of those answers. Just think, if the glass were full to begin with then the pessimist would have had the correct answer. In science, it is imperative to take a realist approach in order to be accurate, rather than to allow one's own character or predisposition to influence one's observations and perceptions.

I have come to realize that *what we believe is irrelevant to the truth, but the truth is very relevant to what we believe.* Beauty enriches the life of every beholder that recognizes it, and so does truth when it is embraced. We cannot change truth; we can merely accept it or reject it. To ignore truth is to reject it. We can call beauty ugly, and say that truth is a lie, but that doesn't make it so.

We can believe what the Bible has to say, or we can reject what it has to say—it is our choice. But that does not mean that it is not relevant to each one of us. It really all depends on whether the Bible is truth or not, doesn't it?

Regarding salvation: if the Bible is false, and you believe it—you lose. If the Bible is true, and you don't believe it—you lose. If the Bible is true, and you do believe it—you still lose—that is, unless you also do what the Bible says you must do. Fortunately, the Bible also says that those who ask shall receive, and those who seek will find, and for those who knock, the door shall be opened; and even better still, it says that with God all things are possible.

It doesn't take a genius to realize that truth is relevant to everybody. Nor does it take a rocket scientist to figure out that truth must be applied in order for its benefits to be realized. But the one obstacle that most people find the most difficult to grasp is this—in a world filled with so many conflicting opinions about truth, how can we really *know* what truth is?

On the next page is an exercise that may seem strange to you, particularly if you are not a born-again Christian. But I ask you to please conduct this exercise with an honest and open heart and mind—no one will even know what you are doing—only you will, so don't worry about what others might think.

The purpose of this exercise, if you have never done anything like this, is for you to experience the power of truth by listening to the voice of your heart, and then by drawing near to God by listening to the still, small voice of Holy Spirit as he bears witness to the truth of God's Word. If you do not have a personal relationship with God already, now would be the perfect time to start one.

As you begin the exercise, just listen to your own heart, and then listen to what God has to say to you, and then do what he says. Remember, *action,* based on truth is what makes the effective difference.

Listening Exercise

STEP 1: Read the following verses from the Bible. After reading them, pause, close your eyes, and listen to the voice of your heart. Hear what your heart is saying to you as you carefully read and focus on what these verses are saying to you. Ask your heart if what you are reading is truth:

"For what shall it profit a man, if he shall gain the whole world, and lose his own soul?"[31] "For he [God] had made him [Jesus] to be sin for us, who knew no sin; that we might be made the righteousness of God in him."[32] "For God so loved the world that he gave his one and only Son, that whoever believes in him shall not perish but have eternal life. For God did not send his Son into the world to condemn the world, but to save the world through him."[33] "And this is the testimony: God has given us eternal life, and this life is in his Son. He who has the Son has life; he who does not have the Son of God does not have life."[34] "The Lord is not slack concerning His promise, as some count slackness, but is longsuffering toward us, not willing that any should perish but that all should come to repentance."[35] "For it is with your heart that you believe and are justified, and it is with your mouth that you confess and are saved."[36] "For it is by grace you have been saved, through faith—and this not from yourselves, it is the gift of God—not by works, so that no one can boast."[37] Jesus said, "I am the way and the truth and the life. No one comes to the Father except through me."[38] "I am the gate;

---

[31] Mark 8:36 (KJV)
[32] 2 Corinthians 5:21
[33] John 3:16-17
[34] 1 John 5:11-12
[35] 2 Peter 3:9 (NKJV)
[36] Romans 10:10 (NIV)
[37] Ephesians 2:8-9 (NIV)
[38] John 14:6 (KJV)

whoever enters through me will be saved." [39] "I am the resurrection and the life. He who believes in me will live, even though he dies;" [40] "... for if you do not believe that I am He, you will die in your sins." [41] "My sheep listen to my voice; I know them, and they follow me. I give them eternal life, and they shall never perish; no one can snatch them out of my hand. My Father, who has given them to me, is greater than all; no one can snatch them out of my Father's hand. I and the Father are one." [42]

STEP 2: Now, try to hear the voice of Holy Spirit—if you seek him with an honest and sincere heart, he will indeed speak to you—pause and listen carefully for his still, small voice within you. Remember, Holy Spirit is the eternal witness, who sees all things from all perspectives. He searches your heart and your mind, and knows you better than you know yourself.

STEP 3: Ask yourself what *you* will do about these words—will you now believe them, and act on them, or will you reject them? The Bible makes it very clear that Jesus is the *only* way to God. God will only accept the righteousness of his beloved son—the blood of Jesus is the *only* sacrifice that will ever be acceptable to God the Father, and it is the *only* offering that can save us from our sins. The Bible says that all men have sinned,

---

[39] John 10:9 (NIV)
[40] John 11:25-26 (NIV)
[41] John 8:24
[42] John 10:27-30 (NIV)

and that all of our righteous acts are as filthy rags to God. Only Jesus was able to please the Father and fulfill the law in order to break law's power over us—and by doing so, he overcame sin and death. He took our sins upon himself and paid the price of sin so that we can be freed from the power of the law, sin and death. He did this to win our love, and to set us free. Galatians 5:1 says, "It is for freedom that Christ has set us free."

Realize that nothing is ever done in the future, and it's too late to do anything in the past. Now is the only time you will ever get anything done. We can only touch the past and the future by the actions we take today. We only live in the present—the now. Ten minutes from now, you either will be saved, or you won't be. You will have either already asked Jesus to come into your heart, or you didn't.

It's not yet too late to repent. If you feel your heart pounding right now, it's because Holy Spirit is stirring within you, confirming the truth. Repentance is turning from *your* way of doing things—from being your own god, and instead, turning toward the one true God, surrendering your life to him so that he can do his perfect work in you, and so that he can teach you *HIS* ways of doing things. Jesus promised that whoever comes to him would not be rejected. He will never leave you, nor forsake you; and that whoever *receives* him becomes a new creation—so that old things are passed away, and all things become new. This is the most important decision based on truth that you will have ever made in your entire lifetime. It can change your life forever—for the good.

When you do, make sure you tell somebody about what you did, and know that because you are not ashamed of him before men, that Jesus will most assuredly confess you before his Father in Heaven—that you are indeed one of his, and that your name will be found written in the Lamb's Book of Life—and Jesus will be with you forever, and you will never be alone again.

Chapter Twelve
What is Man?

As we consider the amazing design of mankind, we may also gain a better understanding of who God is. We are living evidence of whom and what God is like, for he tells us we were fashioned in his image and after his likeness, and it is God's handiwork that we represent. So let's begin with what the Bible has to say about man.

The Man-trinity
Spirit, Soul and Body

The Bible tells us that man is made of three parts: a spirit, a soul, and a body.[43] In order to keep things straight in our minds, let us call these three parts of man the *man-trinity*.

What happens when we compare the man-trinity with the God-trinity? Does this seem at all like an apples-to-apples comparison—body, soul and spirit vs. God the Father, God the Son, and God the Holy Spirit? I think not, because the God-

[43] 1 Thessalonians 5:23 NIV

trinity represents the three persons of God, whereas the man-trinity represents the compositional make-up of a man.

The God-Trinity represents *who* God is, but not necessarily *what* God is. The man-trinity represents *what* man is, but not necessarily *who* man is. So then, comparing the God-Trinity with the man-trinity would really not be a rational comparison at all. So let us then first consider the man-trinity, to see if the compositional makeup of man is anything like the compositional makeup of God.

The Heart and Soul of Man

The living soul was the end product of God's creation of man. For the Bible says[44] that God formed man from the dust of the earth, then breathed into man's nostrils the neshamah[45] (Hebrew, meaning the "breath of life"), and man *became* a living nephesh[46] (Hebrew, meaning "soul").

Since the compositional makeup of a man's soul is not something we can lay before us on a table to study, and since we cannot perceive it with our eyes, let us then study it by considering the many references to the soul of man found in

---

[44] Genesis 2:7 KJV
[45] Strong's no. 05397
[46] Strong's no. 05315

the Bible. In this same way, we will afterward examine the spirit of man to understand how the soul and spirit of man are divided by the word of God. "For the word of God is living and active. Sharper than any double-edged sword, it penetrates even to dividing soul and spirit..."[47]

However, we must not omit the *heart*, because the heart of man is also very important to God, for it seems to be connected with, and is part of, man's soul. We are not referring to the physical heart here, but rather, the spiritual one.

I do not think we need any special consideration of the word *body,* since the meaning is clearly understood already in the context of the man-trinity, for we are already familiar with what the body is physically. So let us begin our exploration of the heart and soul of man, and then we will consider man's spirit.

Our physical heart pumps our lifeblood. Our breathing keeps it going. Stop the breathing and the heart soon fails to function. God breathed into man's nostrils the breath of life, and the newly formed body of man became the whole person of man—a living human being consisting of a soul, a spirit, and a breathing perishable body.

---

[47] Hebrews 4:12a NIV

According to the word of God, the heart of man contains our thoughts and attitudes. Hebrews 4:12 tells us that the word of God judges the thoughts and attitudes of the heart. The heart is the center of feeling and emotion, for the heart can be fearful,[48] sorrowful,[49] glad,[50] proud,[51] obstinate[52] and hateful,[53] to name just a few emotions and attitudes. The heart is also the place of understanding.[54] The heart seems to be the place that holds the intent or motive behind what a person does.[55] It is the place of commitment or lack thereof. The heart seems to be the place of opinion as well. God looks on the heart to see what you really think.[56] The Bible says that God searches our hearts and also our minds, but it seems he is mainly interested in our hearts.

It seems that the heart must be part of the soul, because what is contained in the heart is so closely related to the mind, the emotions, and the will of a person. Also, since the words *heart and soul* are so often found together in the same sentence in so many verses throughout the Bible, it is indicative of an

---

[48] Leviticus 26:36 (NIV)
[49] Romans 9:2 (NIV)
[50] Exodus 4:14 (NIV)
[51] Deuteronomy 8:14 (NIV)
[52] Deuteronomy 2:30 (NIV)
[53] Leviticus 19:17 (NIV)
[54] Deuteronomy 8:5 (NIV)
[55] Proverbs 26:24 (NIV); 1 Chronicles 28:9 (NIV)
[56] 1 Chronicles 28:9 (NIV); 1 Samuel 16:7 (NIV)

inseparable relationship between the two. Therefore, when the soul is discussed, this includes the spiritual heart as well.

2 Corinthians 4:18 (KJV) tells us "…we look not at the things which are seen, but at the things which are not seen: for the things which are seen are temporal; but the things which are not seen are eternal." We all have a physical heart that can be seen, and that is temporal, but we also have a spiritual heart that is unseen and eternal. Lifeblood flows in and out of our physical heart, just as the issues of life flow in and out of our spiritual heart. It flows both ways—in and out. This is an important concept to keep in mind as we consider the heart of man, and see how it relates to the heart of God later on.

The Bible tells us that man consists of a spirit, soul and body.[57] Now it is very commonly taught that the soul of man consists of the mind, the emotions and the will. I could not think of a particular Bible verse that said that so I conducted a Word Usage study[58] of the word *soul* in order to see how the word was used throughout the Bible. It was easy to see why it has become a common teaching. I suggest you also do a Word Usage study as well if you have doubts about this teaching. After observing such a preponderance of evidence of the ways

---

[57] 1 Thessalonians 5:23

[58] See Appendix A at the back of the book for information on how to conduct a Word Usage study.

the word *soul* was used throughout the Bible, I was afterward convinced that the soul does indeed include the mind, the emotions, and the will of an individual, except it seems the soul must additionally contain the spiritual heart as well.

We are already familiar with what a perishable body is, and we have discussed that the soul consists of the mind, the emotions, the will, and the heart. We know from the scriptures that when a man dies, the spirit and soul leave the body,[59] and the body remains on earth and decays. So, in light of what we have learned about the soul, what, then, is the spirit of man?

The Spirit of Man

Perhaps the human spirit may be a little different than generally thought by many. There are a lot of ideas out there about what the human spirit is and where it is located relative to the body. Some people think that it is located around the belly area of the body because of sensations felt in that area, and others think the human spirit is formless and emanates beyond the confines of the body. Some people think that the human spirit should be the part of the man-trinity that rules

[59] 2 Cor. 5:1 "For we know that if the earthly tent we live in is destroyed, we have a building from God, an eternal house in heaven, not built by human hands." See also, 2 Cor. 5:6-10.

over the actions of the soul and body—to put the spirit of man in charge, rather than the body, with its passions and lusts. However, the same passages that many interpret to mean the spirit of man can also be interpreted to mean the Holy Spirit of God instead. Think about it—should man be led by *his* spirit, or by the Holy Spirit? Was Jesus led by his spirit, or by the influence and guidance of the Holy Spirit?

The Bible does not plainly say much about the structure of the human spirit, but there *is* a passage in the Bible that does say that God forms[60] [Heb. *Yatsar*: forms, fashions or frames] the spirit within us. It is the very same word that is used in Genesis 2:7 where "the LORD God *formed* (yatsar) man of the dust of the ground..." It can also be translated "potters," i.e., "God *potters* the spirit within us." The Hebrew word *yatsar* (Strong's No. 03335) is translated into the verb, *potters,* 17 times in the KJV.

So, rather than God giving us a formless or shapeless spirit, this passage would indicate that our spirit is actually formed, fashioned or pottered by Creator God, and therefore it would indicate that the human spirit may actually have a unique spirit-form or shape.

---

[60] Zechariah 12:1 (NIV)

Now the spirit is similar to the soul in that both the spirit and soul are invisible to human eyes. It is our thought-life that makes us who we are as individuals, and *who* we are is contained within our soul, and it is our soul that is responsible to God.

God does not search the spirit of man for accountability, but instead searches the mind and the heart of man—the soul. So what then is the purpose or function of the human spirit?

Does our spirit-form actually have the same shape as our human body, so that when we are absent from the body, we are still recognizable by others after we die? [That is, as long as whoever is observing us has the special ability to see other spirits.]

It is interesting that in Revelation 6:9 (NIV), John writes, "When he opened the fifth seal, I saw under the altar the souls of those who had been slain because of the word of God and the testimony they had maintained." Now you wouldn't think that souls would even be visible since souls are the intangible thoughts, feelings and beliefs of individuals. When I see someone, I recognize him or her by his or her appearance. If I see George, for example, I don't say I saw the body of George or the soul of George; I simply say I saw George. But his name really represents the whole person of who George is, and not just his visible body.

So, for the Apostle John to recognize the souls of those under the altar, he may have been describing them by the way he thought of them, that is to say they must have been visible to him as recognizable individuals, but he also recognized what they had gone through and the difficult sacrifices they had made. People often view groups of victims as souls—"Those poor souls," and therefore John may have described them in this unusual way for that reason.

We know that when a person dies, the soul and the spirit both leave the body, but the body remains and it decomposes. We also know that the resurrection of the righteous dead that Paul speaks about in 1 Thessalonians 4:16-18 has not yet occurred to this day. So, how then could those souls beneath the altar even be recognizable as specific individuals unless they had the appearance or shape of those individuals?

And what about Jesus' account about the rich man and Lazarus in Luke 16:19-32 where he told about the rich man, who was able to look at and to recognize both Abraham and Lazarus despite the fact that they were all dead and separated from their bodies? How could the rich man see, hear, speak and feel without the body parts he had left behind on Earth unless he still retained his bodily shape or form apart from his fleshly, perishable body? And vice versa, how could the rich man also be recognized by Abraham? The rich man must have recognized Abraham and Lazarus as people figures with bodily shapes because the rich man asked Abraham to have his servant Lazarus dip the tip of his finger in water to cool his tongue in order to relieve the torment he was experiencing in Sheol.

Could it be that humans on Earth do not normally see spiritual beings since the surface of the Earth is visible, and therefore it is a perishable, physical environment? And could it be that both Heaven and Sheol are spiritual environments where spirits are readily visible and recognizable by other spiritual beings?

"This is the word of the LORD concerning Israel. The LORD, who stretches out the heavens, who lays the foundation of the earth, and who forms the spirit of man within him…" (Zech.12:1 NIV).

God has amazingly integrated our soul to our body, and it appears he has also integrated our soul to our spirit as well.

MAN

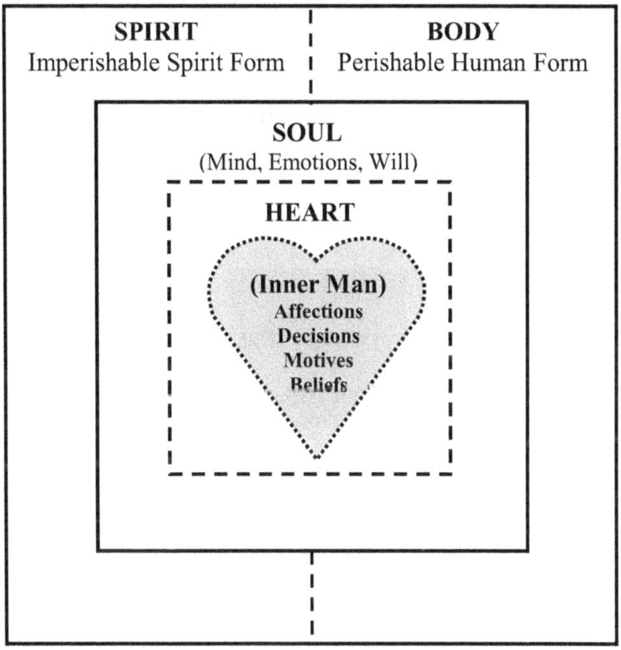

Therefore, it would seem that man consists of the HEART (beliefs, decisions, attitudes, affections, motives, etc.), which is connected to, and part of, the SOUL (mind, emotions, will), which is connected to both the imperishable SPIRIT form, and the perishable human BODY. The believer in Christ will one day exchange his or her perishable physical body for an imperishable physical body for his or her imperishable spirit and soul to inhabit (1 Thessalonians 4:16-18; 1 Corinthians 15:52-54).

The HEART, then, is the innermost core of the SOUL. It is the inner man. It is the treasure chest of the SOUL wherein all the deep-rooted motives, affections, decisions, attitudes, beliefs, morals, and opinions all reside, and which ultimately define the exclusive character and nature of an individual.

The HEART becomes the end product of what the SOUL constructs. That is why the eyes of the Lord focus primarily on the HEART of man, for it is the SOUL that actually produces the HEART.

It seems that the human SPIRIT is the invisible spirit form of an individual; the human BODY is the visible, fleshly form of the same individual. The human SPIRIT and the human BODY occupy the same space.

When the human BODY dies, the human SPIRIT-form still remains with the SOUL intact. Prior to death, the SOUL is connected to and interfaced with both the SPIRIT and the BODY, but when the human BODY dies, the invisible HEART and SOUL still remain intact and interfaced within the human

SPIRIT life-form, thereby preserving the individual's unique form, as well as the character and personality of that individual.

It is important to note also that when New Testament scriptural reference is made to the *spirit* (Greek, pneuma) of a *particular* individual, the soul of that same individual should be included and considered as well, since man's spirit and soul appear to be interconnected. The human soul is integrated with the human spirit much like the human brain is integrated with all the other human body parts, controlling the physical motor movements of the body. The human spirit and soul function together, and together they make up the complete, invisible, eternal, and imperishable spiritual part of man.

But although a New Testament reference to man's *spirit* (pneuma) may include the soul within the context and meaning of the word, most often the Greek word *psychē*, interpreted soul, is used to describe the faculty of man's understanding; but not always, e.g. 1 Cor. 2:11, "For who knows a person's thoughts except their own *spirit* within them? In the same way no one knows the thoughts of God except the Spirit of God."

It seems, then, that the soul is part of the spirit, like the heart is part of the soul. Although each part can be separately described, they are really all part of the overall spiritual configuration of man, designed by God to operate within a physical body that can be perishable or imperishable, temporal or eternal; but the scriptures indicate that the spirit and soul of man still exist even without a physical body.

Chapter Thirteen
Who Is God, and What Is He?

How God describes himself in his written word is important information for us to know. For by learning what the Bible has to say regarding who God is, and what God is, we can then gain a better understanding of who and what we are as well. Just how much *are* we like him, and how are we different?

The God-Trinity
God the Father, God the Son, God the Holy Spirit

The *God-Trinity* or simply, the *Trinity*, is the concept that God is one God, but he is also three distinct persons at the same time: God the Father, God the Son, and God the Holy Spirit, each person of God being distinguishable from the others, but all being unified in purpose. The Hebrew word for God, *'elohiym,*[61] is a masculine noun used in the plural intensive, yet with singular meaning. So, this Hebrew word having been used over 2000 times in the Old Testament is

---

[61] Strong's no. 0430

constantly pointing to the fact that God is a plural, yet singular being.

The Bible, from Genesis to Revelation, has passages that give us clues about who and what God is like. There are a few passages in the English translated versions where God speaks using both singular and plural names and pronouns to represent himself: e.g. "Then God [singular] said, 'Let us [plural] make man in our [plural] image, in our likeness...'"[62] and, "the LORD God [singular] said, "The man has now become like one of us [plural], knowing good and evil…'"[63] and, "Then I heard the voice of the Lord saying, "Whom shall I [singular] send? And who will go for us [plural]?"[64] However, unlike these examples, most other English translated references are singular or plural, but not necessarily singular and plural within the same passage as these are, but in the Hebrew, wherever the word *'elohiym* is used in reference to God, it indicates that God is both a plural and singular God. This is not reflected in most English translations.

God is almighty. He doesn't need man's help to interpret what *he* means to say in His written word. If you simply let God's word speak for itself, He is very clear on this

---

[62] Genesis 1:26 NIV
[63] Genesis 3:22 part
[64] Isaiah 6:8 NIV

issue of the God-Trinity. Just because we have a difficult time understanding a concept, it doesn't mean it is incorrect, and that it is not the truth.

One prime example in the new testament where God manifests himself as three distinct persons, or individuals, is at Jesus' baptism,[65] where God the Father speaks from Heaven confirming Jesus to be his beloved son in whom he is well pleased, and where God the Holy Spirit descends on Jesus in bodily form like a dove, indicating his acknowledgement and favor toward God the Son. And, of course, Jesus himself could be seen, felt, and heard since he was a flesh-and-blood human being. All those that were present at the baptism of Jesus that day, who had eyes to see, and ears to hear, may have had the opportunity to experience with their own senses, on that one occasion, three separate physical manifestations of the Trinity.

When Jesus sent out the disciples, he gave them orders commonly referred to as the Great Commission in which he made reference to the three persons of the Trinity. He said to them, "Therefore go and make disciples of all nations, baptizing them in[a] the name of the Father and of the Son and of the Holy Spirit …"[66]

---

[65] Luke 3:22 (NIV)
[66] Matthew 28: 19 NIV

Jesus often spoke about the Father and about the Spirit of God, but he also represented that he himself was the son of God, such as in Matthew 26:62-64 (NIV) where the high priest said to Jesus, "I charge you under oath by the living God: Tell us if you are the Christ, the Son of God." 64"Yes, it is as you say," [Jesus replied].

Jesus' claim to be the Son of God enraged the chief priests and Pharisees so much; they considered it to be blasphemy. The fact is the Trinity or Tri-unity is very well documented in the scriptures; so let's continue.

The Soul of God

The Bible tells us that man consists of a spirit, soul and body.[67] The Bible also tells us that God has a soul.[68] But since God is also a trinity, the question arises: does each person of that Trinity also have a soul? That is to say, does God the Father, God the Son, and God the Holy Spirit each have an individual mind, individual emotions, and an individual will? Let us consider whether this is so.

---

[67] 1 Thessalonians 5:23

[68] Jeremiah 32:41 (NIV) "I will rejoice in doing them good and will assuredly plant them in this land with all my heart and soul."

Regarding individual minds, the scriptures indicate or infer that the Father has a separate mind from the Son's because Jesus communicated regularly with the Father while Jesus was on the earth. Jesus often got away from the crowds to be alone in order to pray to his Father. If they both had the same mind, there would have been no need for them to communicate in that way.

When Jesus cried out on the cross, "My God, my God, why have you forsaken me?" his mind was clearly separated from the Father's mind.

Jesus was sent and led by the Holy Spirit to go into the desert. Since Jesus followed the guidance of the Holy Spirit, this indicates the Holy Spirit has a separate mind from that of Jesus. The Holy Spirit appears to function individually within the Trinity, but is always unified with the Father and Son in purpose.

Regarding emotions, the Bible instructs us not to grieve the Holy Spirit of God. This indicates that the Holy Spirit has individual emotions, and that he is responsive to how we treat him. Also, at Jesus' baptism the Father demonstrated pleasure and delight in his son, while Jesus' emotions were more in line with showing honor and obedience to his father, for his emotion could not have been delight in a son, since he did not have a son.

Regarding individual wills, in Luke 22:42 (NIV) the Father's will is separate from the will of Jesus as is indicated when Jesus said in the Garden of Gethsemane while praying to the Father, "Father, if you are willing, take this cup from me; yet not my will, but yours be done." Jesus was submissive to the Father's will, doing only those things that pleased the Father, thereby also demonstrating that the Son's will was separate from, but submissive to, the Father's will. Even as Jesus teaches the Lord's Prayer to the crowd, he prays to the Father, "Your will be done on earth as it is in Heaven."

In John 5:21 (NIV), John shows how God the Father and God the Son exercise separate wills according to their pleasure: "For just as the Father raises the dead and gives them life, even so the Son gives life to whom he is pleased to give it."

God the Father was the sending one. God the Son was the sent one. God the Holy Spirit was the enabling one, the supernatural catalyst that not only conceived the virgin birth, but who empowered, equipped and guided Jesus the man to accomplish the will of the Father while on Earth.

The Heart of God

The Bible indicates that God has both a heart and a soul,[69] so if the soul is the mind, the emotions, and the will, what then is the heart of God? As stated already regarding the heart of man, the Bible makes distinctions between the heart and soul by the fact that both words appear together so many times,[70] such as "[with all your] heart and [with all your] soul" within the same sentences, showing separate distinction.

Although the words *heart* and *soul* appear to be used interchangeably in many places, they are not the same thing when considering the heart and soul of an individual. In Hebrews 4:12, the passage clearly tells us that the living word of God judges the thoughts and attitudes of the heart.

The heart seems to be the place of understanding.[71] The heart holds the intent or motive behind what a person does. It is the place of commitment or lack thereof. The heart seems to be the place of opinion as well. God looks on the heart—not the soul—to see what you really think.[72] My mother once said

---

[69] Genesis 8:21; Jeremiah 32:41
[70] Deuteronomy 4:29, 6:5, 10:12, 11:13, 13:13, 26:16, 30:2, 30:6, 30:10; Joshua 23:14; 1 Samuel 14:7; 1 Kings 2:4, 8:48; 1 Chronicles 22:19; 2 Chronicles 6:38, 15:12; Jeremiah 32:41; Matthew 22:37; Luke 10:27
[71] Deuteronomy 8:5
[72] 1 Samuel 16:7

to me, "Your heart is what *you* make of it; your soul is what God made."

The heart is *what* you are. The soul is *who* you are; it is your personal, individual life. The heart contains everything you believe, conclude, and decide; the soul carries out what you do as the result of what you believe and decide in your heart. The heart holds the motive; the soul commits the crime so to speak. The heart harbors the attitude, i.e. hatred, and then decides to kill; the soul commits the murder, and, according to the Bible, it is the soul that sins.

The heart conceives what the soul performs via its interface with the body, and the soul who sins is therefore judged for the crime. In Ezekiel 18:4 (NIV), God declares, "For every living soul belongs to me, the father as well as the son—both alike belong to me. The soul who sins is the one who will die."

Unlike the heart of man, the scriptures seem to indicate that God, as three persons (God the Father, God the Son, and God the Holy Spirit), has a common heart. The heart of the Father, the Son and the Holy Spirit seems to be one and the same. To best describe the heart of God, we must first consider that while the Father, the Son and the Holy Spirit each have individual minds, emotions, and wills, i.e. individual souls, the scriptures also speak of God's soul,[73] and indicate that God's soul is a collective or triune soul, one that consists of the sum-total of the individual souls or persons of the Godhead—the separate minds, separate emotions and separate wills, all converging into a single, triune soul.

And it is this triune soul that then produces the unified and singular heart of God. For God's heart is the place where the three persons of the Trinity are truly only ONE God. And God wants your heart and my heart to become one and to beat as one with his heart. Jesus was our perfect example of how this can be accomplished as a human being. Jesus always did those things that pleased the Father. In order for us to emulate Jesus, Paul, in Philippians 2:5 (KJV) says, "Let this mind be in you, which was also in Christ Jesus…" Indeed, it must require the mind of Christ to develop the heart of God within us.

Although there is some indication that the heart of God may appear to be individualized, in actuality the singular heart of God simply manifests outward from each person of the Trinity. Remember, things flow into the heart, and things flow out from the heart. Matthew 15:18-19 (KJV) says, "[18]But those things which proceed out of the mouth come forth from the

---

[73] Psalm 11:5 (NIV) "The LORD examines the righteous, but the wicked and those who love violence his soul hates.

Jeremiah 32:41 (NIV) "I will rejoice in doing them good and will assuredly plant them in this land with all my heart and soul."

heart; and they defile the man. [19]For out of the heart proceed evil thoughts, murders, adulteries, fornications, thefts, false witness, blasphemies…" And in Proverbs 4:23 (NIV), the Bible says, "Above all else, guard your heart, for it is the wellspring of life."

Again, the individual souls of the God-Trinity converge to form the triune soul of God, which then produces the single heart of God, which in turn emanates and manifests from Father God, Jesus and Holy Spirit. The figure on the next page illustrates this.

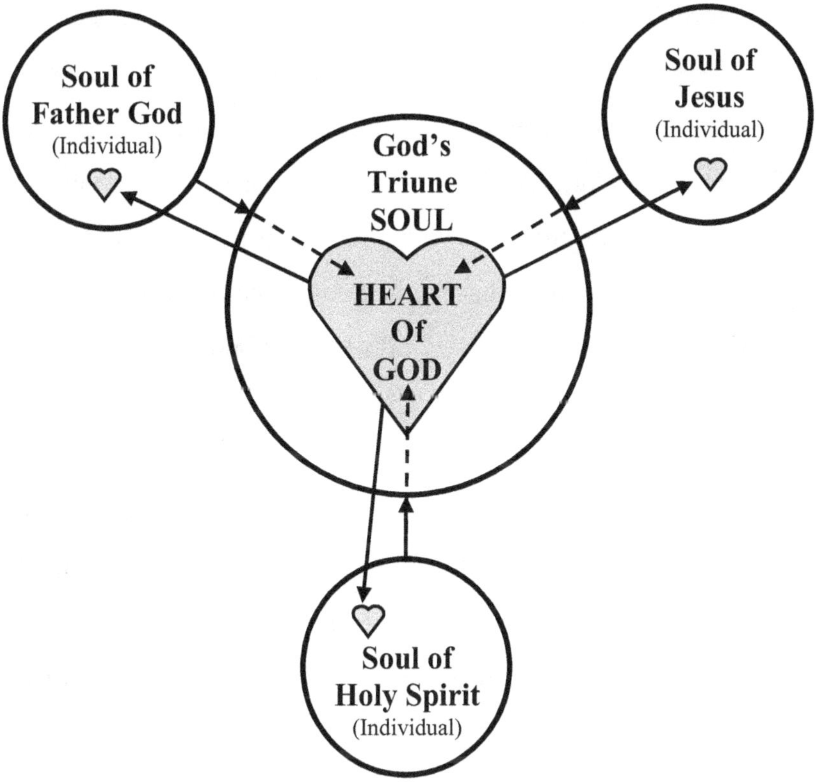
Soul of
Father God
(Individual)
Soul of
Jesus
(Individual)
God's
Triune
SOUL
HEART
Of
GOD
Soul of
Holy Spirit
(Individual)

God the Father

The Bible both implicitly and explicitly states that God has bodily form like man. Now whether this bodily form is a spirit form, and a physical, imperishable form, it is not clear. The Bible *does* say that we are created in his image and after his likeness. The Bible describes him as walking in the Garden of Eden near Adam and Eve.[74] God was partially observed by Moses at Mount Sinai,[75] where Moses asked to see God's glory. God told Moses that no man could see his face and live. But God placed Moses in the cleft of a rock and covered Moses with his hand as God's goodness passed by. Then God removed his hand so Moses could see his back.

Doesn't it seem like God the father's body must be very large if he could actually cover Moses with his hand—especially since that hand could also lift Moses off his feet and place him into the cleft of the rock for protection? Doesn't it stand to reason that since God created man in his own image and after his likeness, that man's body should appear similar to God's?

Revelation 5:7 (NIV) says, "He [the Lamb that had been slain] came and took the scroll from the right hand of him

---

[74] Genesis 3:8
[75] Exodus 33:19-21

who sat on the throne [the Father]. This passage describes God the Father as having a right hand. It would be reasonable to rationalize that he probably has a left hand as well, and two arms to go with those hands.

The Bible also says that God the father is seated in heaven on the throne, and that Jesus is seated at the right hand of the Father, so both God the Father and God the Son are at specific locations in heaven. God the Father does not occupy the same space as God the Son, for Jesus is seated at the right hand of the Father, and therefore they also must have separate bodies as well. It is unclear if God the Father has just a spirit body form, or if he also has a physical, imperishable body form.

God the Son

The first chapter of the Book of St. John reveals the following about the living Word of God, obviously being Jesus, the *only* begotten Son of God:

> [1]In the beginning was the Word, and the Word was with God, and the Word was God. [2]He was with God in the beginning. [3]Through him all things were made; without him nothing was made that has been made…
>
> [10]He was in the world, and though the world was made through him, the world did not

> recognize him. [11]He came to that which was his
> own, but his own did not receive him. [12]Yet to all
> who received him, to those who believed in his
> name, he gave the right to become children of
> God— [13]children born not of natural descent, nor
> of human decision or a husband's will, but born of
> God.
>
> [14]The Word became flesh and made his dwelling among us. We have seen his glory, the glory of the One and Only, who came from the Father, full of grace and truth.

Therefore, from these passages we understand that Jesus was indeed God, and that he was with God even in the beginning [indicating a singular but plural God], and that he was a co-Creator of Heaven and Earth.

Verse 14 tells about the Word becoming flesh, but what bodily form did the Word have before he became flesh and blood? Whether Jesus had just a spirit body form at the time of creation, or whether he also had a physical body form that was imperishable, similar to his resurrection body, the Bible is unclear about this. The Word only references when he became a human being, when he took on flesh.

While on the earth, Jesus had a perishable, physical body form that was once transfigured temporarily, and once died, but it was his physical body form that was then resurrected from the grave as the firstborn from among the dead, resurrected and transformed into an *imperishable*, physical body form. Yes, Jesus was the firstborn from among the dead to be resurrected into an imperishable, physical body form, because all others who had been resurrected from the dead before him, such as Lazarus, were resurrected only to return to their same perishable physical bodies, which eventually died again. Also in Matthew 27:50-53, after Jesus was crucified and gave up his spirit, it says that the earth shook, the rocks split, and the tombs broke open. It says that the bodies of many holy people who had died were raised to life, and that they came out of the tombs, went into the holy city and appeared to many people. Think of it—these all got to experience death twice—one time would be enough for me.

Jesus was the firstborn from among the dead, resurrected and transformed into an imperishable physical body form that could be touched and felt. What kind of body form do you think Moses and Elijah had on the Mount of Transfiguration where they were both observed by Peter, James and John, speaking with Jesus?

Also, consider the following verses in Colossians 1:15-18 (NIV) that say this about Jesus:

> [15]He is the image of the invisible God, the firstborn over all creation. [16]For by him all things

> were created: things in heaven and on earth, visible and invisible, whether thrones or powers or rulers or authorities; all things were created by him and for him. [17]He is before all things, and in him all things hold together. [18]And he is the head of the body, the church; he is the beginning and the firstborn from among the dead, so that in everything he might have the supremacy.

God the Holy Spirit

The Holy Spirit is the third person of the Trinity who is filled with God's knowledge, understanding, wisdom, counsel, power, and the fear of the Lord. Unlike God the Father and God the Son, the Holy Spirit has no particular bodily form, but instead has the ability to manifest in unlimited forms such as when he descended upon Jesus in a bodily form appearing as a dove at Jesus' baptism.[76]

Of the many things found in the Bible regarding the person of the Holy Spirit, these are only some of the things Holy Spirit does:

---

[76] Luke 3:22

Holy Spirit hovers over,[77] indwells,[78] rests on,[79] comes on,[80] fills,[81] baptizes,[82] empowers and strengthens,[83] enlightens and reveals,[84] instructs and teaches,[85] counsels,[86] moves,[87] leads,[88] convicts,[89] gives special abilities for ministry,[90] speaks through people,[91] admonishes,[92] creates,[93] gives rest,[94] conceives and gives life,[95] lifts up,[96] gives visions,[97] drives out demons,[98] prophesies and testifies,[99] fills with joy,[100] gives

---

[77] Hovers over: Genesis 1:2
[78] Indwells: Ezekiel 36:27
[79] Rests on: Numbers 11:25; Isaiah 11:12
[80] Comes on: Numbers 24:2; Judges 6:34;
[81] Fills: Micah 3:8; Luke 1:15; Ephesians 5:18
[82] Baptizes: Mark 1:8; Luke 3:16; 1 Corinthians 12:13
[83] Empowers and Strengthens: Judges 14:6; 14:19; 1 Samuel 10:10; 1 Samuel 16:13: Luke 4:14; Acts 1:8; Acts 9:31
[84] Enlightens and Reveals: Luke 2:26; 1 Corinthians 2:10; Ephesians 3:5
[85] Instructs or teaches: Nehemiah 9:20; Luke 12:12; 1 Corinthians 2:13; Hebrews 10:15
[86] Counsels: John 14:26; 15:26 [Note that the KJV translates the word "counselor" to mean "comforter" instead.]
[87] Moves: Ezekiel 36:27; 37:1
[88] Leads: Matthew 4:1; Romans 8:14
[89] Convicts: John 16:7-11
[90] Gives gifts to minister: 1 Corinthians 12:8-10
[91] Speaks through people: 2 Samuel 23:2; Acts 2:4; 2:18; 6:10
[92] Admonishes: Nehemiah 9:30
[93] Creates: Psalm 104:30
[94] Gives rest: Isaiah 63:14
[95] Gives life: Matthew 1:18; 1:20; John 3:6; 6:63; 2 Corinthians 3:6; 1 Peter 3:18
[96] Lifts up: Ezekiel 3:12; 8:3; 43:5
[97] Gives visions: Ezekiel 11:24
[98] Drives out demons: Matthew 12:28
[99] Prophesies and Testifies: Luke 1:67; 1 Timothy 4:1; 2 Peter 1:21; 1 John 5:6

gifts,[101] encourages,[102] compels,[103] sends us,[104] intercedes,[105] sanctifies,[106] justifies and seals us,[107] ministers,[108] frees,[109] grieves,[110] gives help,[111] helps us worship,[112] helps us pray,[113]and renews us.[114]

Psalm 139: 7-10 (NIV) enlightens us about the Omnipresence of the Holy Spirit: "Where can I go from your Spirit? Where can I flee from your presence? [8] If I go up to the heavens, you are there; if I make my bed in the depths [Sheol], you are there. [9] If I rise on the wings of the dawn, if I settle on the far side of the sea, [10] even there your hand will guide me, your right hand will hold me fast."

Proverbs 15:3 (NIV) says, "The eyes of the LORD are everywhere, keeping watch on the wicked and the good."

---

[100] Fills with joy: Luke 10:21; 1 Thessalonians 1:6
[101] Gives gifts for ministry: Acts 2:38; Hebrews 2:4
[102] Encourages: Acts 9:31
[103] Compels: Acts 20:22
[104] Sends us: Acts 13:4
[105] Intercedes: Romans 8:26-27
[106] Sanctifies: Romans 15:16; 2 Thessalonians 2:13; 1 Peter 1:2
[107] Justifies and seals us: 1 Corinthians 6:11; Ephesians 4:30
[108] Ministers: 2 Corinthians 3:8
[109] Frees: 2 Corinthians 3:17
[110] Grieves: Ephesians 4:30
[111] Gives help: Philippians 1:19; 2 Timothy 1:14
[112] Helps us worship: Philippians 3:3
[113] Helps us pray: Jude 1:26; Romans 8:26
[114] Renews us: Titus 3:5

Jeremiah 23:23-24 (NIV) says, "Am I only a God nearby, declares the LORD, and not a God far away? Can anyone hide in secret places so that I cannot see him? declares the LORD. Do not I fill heaven and earth? declares the LORD.

The Holy Spirit seems to be the omnipresence of God, the fabric of the universe that fills all space throughout the vast cosmos, including all submicroscopic and subatomic space.

Because God is a Trinity, God the Father, God the Son, and God the Holy Spirit, each person of the Trinity contributes to God's triune soul, and to the single, unified heart of God.

The following God model shows the omnipresence of the Holy Spirit, and the relative locations, and bodily forms of the Trinity, and of man prior to man's death:

God & Man Model

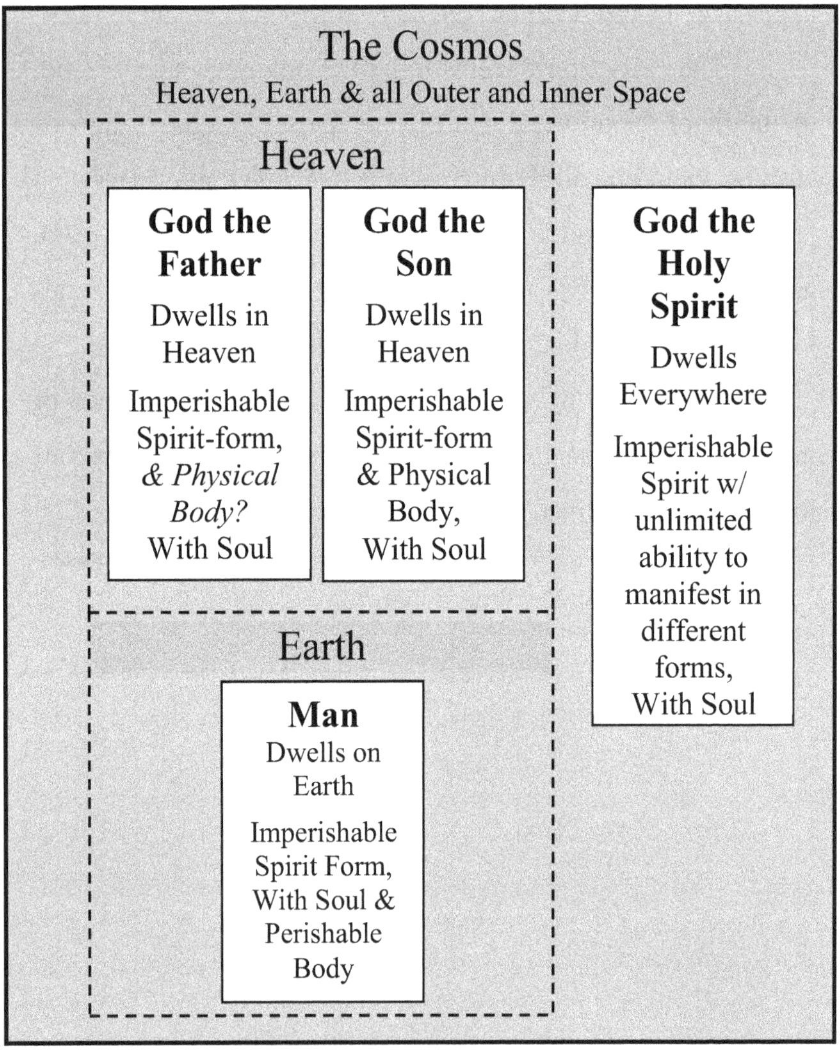
The Cosmos
Heaven, Earth & all Outer and Inner Space
Heaven
God the Father
Dwells in Heaven
Imperishable Spirit-form, & Physical Body? With Soul
God the Son
Dwells in Heaven
Imperishable Spirit-form & Physical Body, With Soul
God the Holy Spirit
Dwells Everywhere
Imperishable Spirit w/ unlimited ability to manifest in different forms, With Soul
Earth
Man
Dwells on Earth
Imperishable Spirit Form, With Soul & Perishable Body

Part III: Truth-Science Methodology

Chapter Fourteen

Signature & Fingerprints of God

A symphony is recognized by its rhythms, themes and motifs. A seasoned composer is recognized by his or her unique style of composition, and by the way the instruments have been skillfully or uniquely grouped and voiced. It is similar for professional painters, instrumentalists, authors, poets and vocal artists – they have uniqueness. All master artists have gone beyond the mimicking and the cliché, to the point of having developed and demonstrated their own styles.

As a student of music, I had to train myself to recognize the works of many composers. To train our ears, my fellow students and I would go into the music library and check out many recordings by specific composers. We would then randomly listen to short portions of many of their works and discuss what we heard in order to identify specific qualities or characteristics of the music that were uniquely theirs alone, in order to identify their signatures, so to speak. After a while, we would begin to recognize their music even when we had not heard the particular piece before.

The more time spent listening to any one composer's works, the easier it was to recognize that composer. It required time to develop the skill, but it was necessary in order to pass the "guess the composer" test in Music History class. Likewise, the more time a person spends studying and knowing what the Bible says, the more discerning that person will become, and the more quickly that person will learn to recognize God's style.

Fingerprints are unique. It's easy to see a greasy auto mechanic's fingerprints on a sky blue colored classic car fender, but most of the time fingerprints are not even visible to the naked eye. Detectives have developed various techniques for discovering and preserving otherwise invisible fingerprints. Revealed fingerprints are helpful in determining who may have handled an object used in a particular crime. Those fingerprints, when skillfully lifted and preserved, also serve as evidence or proof for the prosecution of an accused suspect. Just think— something invisible is made visible— something hidden is revealed.

God's style and signature are everywhere, and so are God's fingerprints. It is easy to recognize his style and signature, but his fingerprints are more difficult to discern or to see. They are generally invisible at first.

I once had a boss that had a handwriting style that was very recognizable as his handwriting—everyone in the office knew who wrote it; you could even go so far as to call his writing style his signature, even though he didn't actually sign it, because it was so uniquely his handwriting. Those outside the office would not recognize the handwriting, and wouldn't know who wrote it, because they were unfamiliar with his style. It took a long time, but eventually, those close to him developed the skills necessary to actually read readily what he had written.

God's signature is the same way. It is easy to spot by those who are willing to recognize and acknowledge his existence, just as it is easy to recognize the Pablo Picasso style after you have seen a sampling of his paintings. But it takes a skilled professional who has spent much time studying Picasso's art to recognize the originals from the better copies. And God's fingerprints are often very difficult to discern, and his words are often difficult to understand except by those who have taken the time to know God, to know his style, and to learn to hear the still, small voice of Holy Spirit as he reveals his ways to us in order that we may interpret and appreciate his creative handiwork.

Anything that a master artist creates will have certain qualities, characteristics, motifs, or attributes that are typical and indicative of the artist's style. God the Creator has a style all his own, for sure. A few of these more general characteristics will be discussed so that they will be easier to recognize. Someone who has been trained to appraise antiques can quickly recognize the genuine among all the fakes. So as we learn some of the qualities, characteristics, motifs, and attributes that represent the unique technique of the Creator of all things, we can then use this knowledge and understanding to help us unfold or unlock the mysteries of whatever aspect of God's creation we are studying.

So then, God's *signature* or *style* consists of the more obvious qualities of his creation—those features that are readily visible to those who merely acknowledge God's existence—for those with eyes to see, these things can be observed and will not be invisible. These more obvious things serve to direct or point us toward the *knowledge* of God's creation. But God's *Fingerprints* are the more hidden characteristics, motifs, and attributes woven into his creative handiwork that help or serve to guide us even further to an *understanding* of the inner workings of his creation or his design. Remember the Bible tells us to gain knowledge, then gain understanding. Finally, gain wisdom. Wisdom is the ability to skillfully apply the knowledge and understanding gained in order to achieve something useful and productive.

Following are some of the general characteristics of God's style—God's writing style and signature. A few examples of God's fingerprints will be presented later on. And as we become more sensitive to God's ways, we will also become more streamlined and successful in discovering and unlocking the mysteries of God's creation.

But remember, the journey begins with an honest quest for the truth, a friendship relationship with God, a heart of faith, with eyes ready to see, and ears expecting to hear the still, small voice of Holy Spirit as he interacts with our heart, mind and soul.

Following are introductory examples of how to recognize God's signature in creation, with emphasis on the word *introductory*:

Functions and Cycles

The Bible says, "To everything there is a season, and a time for every matter or purpose under heaven" (Ecclesiastes

3:1, Amplified Bible). Because we know that everything God created has a season and a time when its purpose is fulfilled, we know that everything God made has a purpose, and we also know there are cycles we should be looking for. Everything has a function and a cycle. Everything exists for a reason and a season. *Look for cycles and purpose in everything God has created.*

Layered Reality

Theoretical physicists have been trying for decades to find a unified theory that will tie together Quantum Mechanics with General Relativity – the micro to the macro. Many millions of dollars and countless hours have been devoted to this endeavor. Why is the marriage between the two so important? Where is the justification, and why does String Theory seem to keep recycling again and again through various science books and magazines?

Quantum Mechanics and General Relativity are already unified in that they are actually two perspectives of the same universe. General Relativity deals with the macro universe or the bigger picture of large bodies in motion. Quantum Mechanics deals with the micro universe, the movement of subatomic particles within an atom. Both are useful, as long as each method is applied appropriately.

If we are to understand the world around us more thoroughly, we must perceive it at many different levels or layers. We don't see the more complete picture unless we consider as many layers as possible or perspective truths. For example, in Psychology there is the person others perceive you to be, the person you believe others perceive you to be, the person you believe you are, the person you were, the person you want to be, and the person you really are that you are trying to understand, just to give a few examples.

We are all very layered both physically as well as psychologically. There is the social you, and there is the uninhibited you that manifests when you are alone. There is the you at work, and there is the you at home. Each of these layers or perspectives may be considered and investigated based on different dynamics and by using different technologies and methodologies. Each perspective of you is still you. A medium-range photograph of you may easily be recognizable by others, whereas a close-up photograph of your fingerprint may not, yet your fingerprint is identifiably you as is the medium-range photograph, although the medium-range perspective reveals more identifying information about you that others would more easily recognize.

There is also the dynamic aspect of who we are both physically and psychologically because we are always changing. We can monitor the person we were at different time intervals to get a picture of who we are becoming. We can look at the trends in our behaviors of the past in order to learn if we are progressing or digressing. We can also take snapshots of our appearance at various time intervals to see how we are changing physically.

When we use a microscope to view a sample of a leaf, we use different powered lenses to view different depths and aspects of the same sample, and what is viewed appears distinctly different by changing the power of the lens. One perspective may reveal the patterned arrangement of the cells that make up the epidermal layer of the leaf, while another perspective using a higher powered lens may reveal the tiny chloroplasts spinning around within the protoplasm inside an individual cell as they transform light energy into chemical energy during the process of photosynthesis.

We use mathematics that are layered into many disciplines, higher and lower such as Calculus, Trigonometry, Algebra, Geometry, Multiplication, Division, Subtraction,

Addition, and so forth to help us understand and accomplish so many tasks within our daily life experiences on Earth. For help in space exploration, and to better understand the universe and what happens to objects in outer space, we have the theory and mathematics of General Relativity. For the invisible, atomic and subatomic world we have the theory and mathematics of Quantum Mechanics to help us understand what we cannot naturally perceive through our senses. These particular disciplines, like the different powered lenses for the microscope, enable us to perceive the recognizable world around us as well as those layers that exist that we would not normally be able to perceive with our eyes, ears or other senses.

Layered Reality is a realistic, practical approach that incorporates gathering as many perspective truths as possible in order to see the bigger picture. It is good science. Layers separated by intervals are natural phenomena everywhere we look.

In Chemistry, the electrons of each chemical element are ordered into distinct levels and sublevels separated by intervals. This is true with the table of the chemical elements as well. In music, a single fundamental tone or frequency is made up of many layered tones known as harmonics that are each separated by particular intervals. Light can be subdivided into the many frequencies of the light-wave spectrum, and even the physical nature of light itself is layered in that it may be observed as both particle and wave.

The atmosphere is layered. The soil is layered. The ocean is layered. The sky is layered. Skin is layered. The rain forests are layered. The Bible is layered. With so many things layered, we would probably be better off learning how to look at and work within the layers instead of spending millions of dollars trying to tie them together. *Expect everything God has created to have many layers or perspectives of understanding and purpose.*

Contrast, Repetition, Sequence, & Symmetry

When my son was born, I was filled with the typical hopes, dreams and fears that any father would have for his first-born child. I wanted him to be healthy, happy and successful. As I watched him grow, I was amazed at how quickly he developed patterns of behavior and a certain uncanny ability to control those around him. Have you ever noticed the mysterious ability infants have to alter adult behavior? The next time you're at the mall, just watch what happens to adults when they gather around a cute baby. The baby suddenly becomes very adult-like as it gazes back with a serious look at the strange way the adults are talking and behaving— it's an amazing role reversal. The adults are all in the child's face making weird expressions and talking baby talk.

My wife and I had all we could do to keep up with our curious and eager little boy. Our lives suddenly changed, and I found myself doing things I would never have imagined myself ever doing before. I found myself carrying things I never would have carried, and my gag reflex was tested both day and night.

I remember observing him as he lay belly down on the sofa beside me. His small, grapefruit-sized head quivered as he tried lifting it to new heights in an attempt to see all the interesting things around the room. He looked so helpless – so needy – so innocent – but mostly he appeared curious. As his head turned awkwardly toward me, our eyes briefly engaged, and I smiled a silly smile to get a reaction from him. But all I could see was the frustration in his face. I wondered what it would be like not to have language at a time when you were being exposed to so many new, colorful and exciting things. What would it be like not to know what things are, or not to know why you feel the way you do. My heart was filled with compassion for him. I wondered what I could do to help him more quickly learn the names of things, and to help him understand the complex world that surrounded him.

I decided to speak the name of whatever his attention was focused on. For example, when he saw the tennis ball I held in front of his face, I said the word ball, and so on. But how could I help give him secondary knowledge about things? In other words, how could I help him to understand that the object was not only a ball, but it was a green ball as well, and it was also a fuzzy ball, and that the ball also bounces up and down?

As I pondered the most effective way to do it, it eventually occurred to me that contrast must be the answer! How could he know what a green ball was if he didn't know what a red ball was, or a yellow ball? How could he know that the ball bounces up and down if he didn't know what up and down were?

So once my son understood, "ball," I could then teach him to know all his basic colors as well, just from that one object by using contrast: "blue ball," "red ball," "yellow ball." And to really anchor the concept in his mind, I could then hold another object in front of his face, "car." Then remove it and replace it with, "ball." Next would be, "Blue ball," "Blue car," and so on.

I shared these brilliant ideas with my wife, whose education was in the field of Early Childhood Development. She told me there were already books available with contrasting things like that for young children so we purchased some, used them, and we ended up with a little boy who hasn't stopped talking since, and who at six months could say, "More geen beans, pwees."

So we were able to teach our son the names of objects by walking him around and pointing to things while saying the names, or else by holding the object in front of him and saying its name. By repeatedly doing this, the repetition and consistency reinforced the knowledge.

By showing him contrasting things, he was able to understand the concept of differences such as: cold water vs. warm water – up vs. down – light vs. dark – big (large) vs. little (small), and so on. We taught him sequences like small, medium, and large. By teaching him his body parts and many descriptive and contrasting words, he was able, at a very young age, to tell us specifically how he was hurting and how he felt, so we could better understand how to help him.

So then, *contrast, accompanied by patterns of repetition and sequence, are the building blocks of basic human knowledge and understanding.* More abstract concepts such as beautiful vs. ugly can be understood by considering patterns and symmetry vs. the lack thereof, or more simply, the contrast of having or having not.

These same basic learning principles are useful in science, when exploring unknown phenomena or situations. *Since God was orderly and artful in everything he created, then we should expect to find, and we should be looking for, the built-in order that is inherent in whatever it is that we are studying. We should be looking for contrast, patterns of repetition, sequences and symmetry, all while considering that everything has purpose or function.*

A good knowledge of the various mathematical disciplines will also equip an individual to recognize the intrinsic mathematical patterns that are part of God's creation that surrounds us. Look for patterns involving the squares or cubes of numbers, numerical symmetry, and functions. For example, many solutions can be broken down to an A = B x C concept such as Distance = Rate x Time, or Area = Length x Width, and so on.

Concerning symmetry, it has been my experience that when God makes something symmetrical, the symmetry is generally complete. In other words, the right side of a person's face is not perfectly symmetrical with the left side, but there are still two complete sides to the face that are mostly symmetrical. Very seldom do you find perfect mirror symmetry, right and left, or top and bottom, but it is most often complete symmetry, unless it is abnormal. So, *hidden symmetry is something we should pay attention to also, because it is a fingerprint of the Creator.*

Primary and Secondary Patterns

Another fingerprint example is this: Sometimes when things have been organized into a particular logic-based arrangement, secondary patterns or geometric shapes form as well. These secondary patterns or shapes only exist as the conditional result of the primary arrangement. Secondary patterns can be other types of sequences, geometrical shapes, or even new logical ideas or concepts that are the result of the primary pattern or arrangement.

In 2005, while eating lunch at the local Burger King, I decided early that week that during my lunch hour I would study the periodic table of the chemical elements. I brought with me a periodic table from the Internet that I had downloaded and printed out the night before [It is not shown here for copyright reasons].

As I looked at the periodic table, I quietly asked God to show me his elements. As I observed the complex table before me, I began noticing many numeric sequences within the atomic numbers and also some symmetry within what I eventually learned were called *electron shell configurations.* I observed a symmetrical pattern in the shell configurations of the Alkaline Earth Metals group that drew my curiosity.

There was unique symmetry within that group that didn't exist in any of the other element columns, and the symmetry was sequential. I also noticed it was incomplete, ending with the element called Radium (Ra), atomic number 88. [See the figure showing the Alkaline Earth Metals group on the next page].

So I then decided that I would complete the *single-double pattern* I had observed, and this would complete the symmetry for these electron shells [They are the numbers the arrows are pointing toward in the figure].

After doing so, I realized that with the pattern and symmetry completed, there had to be a missing element that would have 120 electrons, element 120.

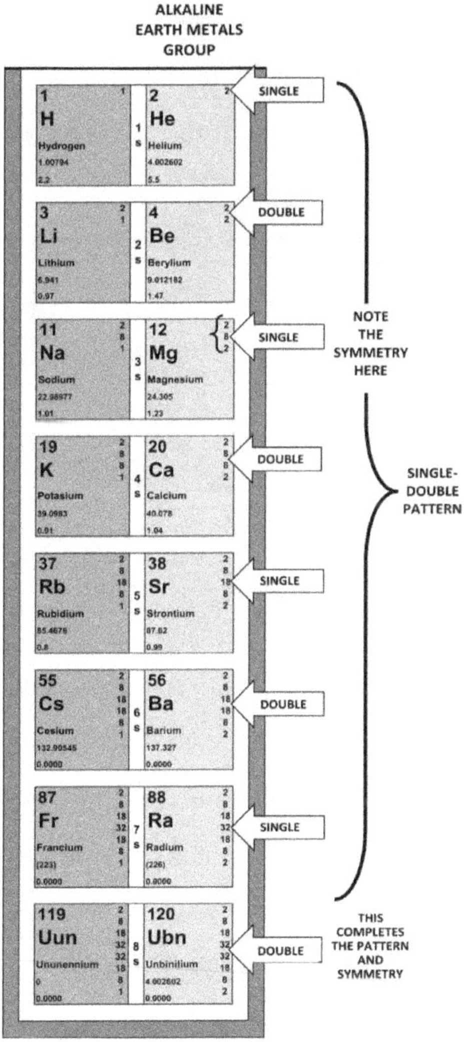

Shortly after that, another idea popped into my mind. [I believe it was a direct answer to my prayer of asking God to show me his elements.] The idea came in the form of a directive: "Keep this Alkaline Earth Metals column together, but move it so that element 120 is at the bottom right-hand corner of the element table."

So when I got home from work that evening, I reordered the elements using a spreadsheet so that element 120 would be located at the bottom right-hand corner of the table. I then rebuilt the element table based on element 120, keeping the Alkaline Earth Metals group intact, and to the far right-hand side of the table. I reordered the elements by atomic number sequence to the left of that column, and what happened was astonishing. I had never seen the elements in such a simplified arrangement before:

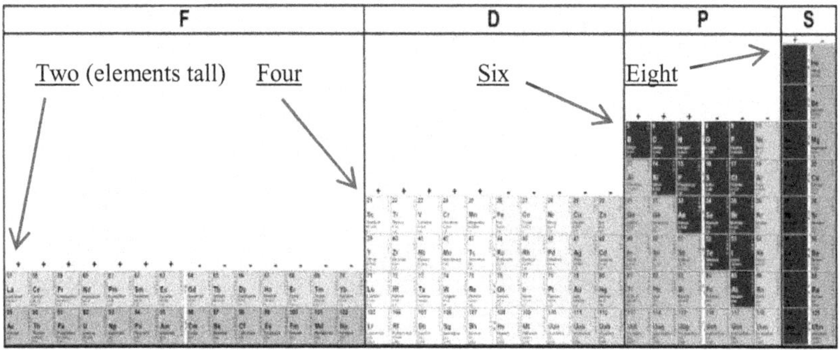

Chemical Block Table of the Elements
Copyright MMV by Steven L. Signor

I originally placed elements 1 and 2 (Hydrogen and Helium) at the top of the third and fourth column from the right, respectively, so that Helium would be above the other Noble Gases elements—but that didn't fit the pattern I was observing, where each element block was increasing in height by two elements (looking at the table from left to right), so I moved them instead to the top of the S block to complete the pattern. The elements were now orderly within their groups and element blocks, but Helium was not with the other Nobel Gases. I had to abandon what I had been taught by modern day science—that Helium, element 2, belongs at the top of the Noble Gases group.

The element blocks (F, D, P & S), now appearing orderly, were aligned into perfect numeric patterns, with subshell fill sequences aligning as well, and each block resembling a rectangular shape. The number or quantity of elements forming the height of each element block appearing in a sequential pattern (from left to right) of 2, 4, 6, and 8, or 2 x **1** = 2, 2 x **2** = 4, 2 x **3** = 6, and 2 x **4** = 8; and the quantity of elements contributing to the four widths of rows of the table (beginning at the top and going downward) forming a 2, 8, 18, 32 pattern that breaks down to 2 x $\mathbf{1}^2$ = 2, 2 x $\mathbf{2}^2$ = 8, 2 x $\mathbf{3}^2$= 18, and 2 x $\mathbf{4}^2$ = 32. Observe the emboldened numbers above.

Many secondary patterns had resulted because of the new arrangement of the table, and these secondary patterns provided an impressive mathematical confirmation that demonstrated the new arrangement to be the correct one, the way God had designed it to be.

Notice also that a geometric shape also occurred with the whole element table as well—a right-triangular shape. And although the hypotenuse of the right-triangle is distorted, a right-triangular shape still best describes the general shape of the new table, whereas the shape of the conventionally accepted periodic table does not resemble any standard geometric shape due to the misplacement of Helium.

So, when the chemical elements are aligned in this more logical manner, we find that Helium does not align vertically with the other Noble Gases group elements as conventionally taught. It aligns instead with the Alkaline Earth Metals group. Moreover, the relocation of Helium is what completes the design of the right-triangular geometric shape of the whole table, and its relocation also causes the P and S element blocks to appear rectangular in shape, removing the extra element from the P block, and adding the missing element to the S block. Moving Helium was like placing the final puzzle piece in order to complete the picture puzzle.

So, it appears that the Noble Gases *group* is really a misnomer—as far as its being referred to as a single column group anyway. After all, just because elements behave similarly, it does not always mean they must belong in the same column in the element table. Iron, Cobalt, and Nickel each have magnetic properties, yet they are not found within the same column of elements either.

Perhaps the reason Helium functions as a Noble Gas is because it is located at the end of its energy level or period—just like the other Noble Gases are located at the ends of their respective energy levels.

When considering the geometric shape of the Block Table of the Chemical Elements, it *only* makes sense when Helium is at the top of the Alkaline Earth Metals column; otherwise, the geometric shape formed by the rest of the elements remains distorted. [I might add also that the 120 chemical elements can be arranged to form an eighteen-sided, three-dimensional crystal too, but the perfect crystal is *only* formed when Helium is placed properly at the top of the Alkaline Earth metals group]. I know this sounds complicated, but it really wasn't. It was fun and interesting.

The first step was to ask God for wisdom and insight, which I did—I asked God to show me his elements. The second step was to look for logical patterns and sequences in order to better understand what appeared complicated. The third step was to complete any patterns that appeared to be incomplete. After this, the new arrangement was carefully looked at, and secondary patterns and geometric shapes were observed that had developed as the result of new arrangement.

Afterward, I asked God questions about it. Eventually, he answered all of my questions—some answers came much sooner than others, but in time God answered all of my questions. Sometimes God needed to teach me other things so I could understand the answers to my own questions.

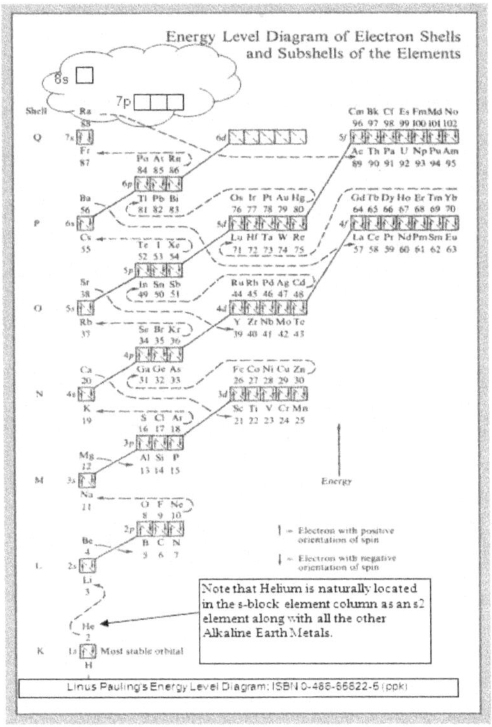

It is also interesting to note that the Block Table of the Chemical Elements is naturally in perfect subshell fill sequence, a sequence taught and illustrated by Linus Pauling in his *Energy Level Diagram of Electron Shells and Subshells of the Elements* that may be found in his General Chemistry book[115] (see the figure above).

By adding the additional elements up to atomic number 120 (clouded at the upper left-hand corner), it completes the

[115] General Chemistry, Linus Pauling's Energy Level Diagram: ISBN 0-486-65622-5 (pbk)

illustration that Mr. Pauling used. Note that Mr. Pauling's diagram is essentially the Block Table of the Chemical Elements turned upside down, except that all the elements have been doubled up within the blocks in Mr. Pauling's diagram.

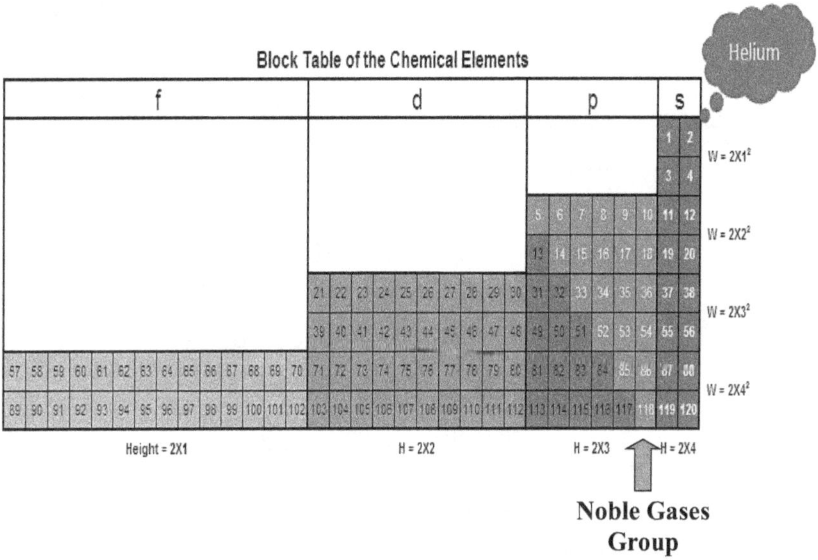

So, the subshell fill sequence is yet another secondary pattern that naturally forms when all of the elements are reorganized into the arrangement of the *Block Table of the Chemical Elements* (shown above).

An arrangement of the more traditional periodic table is shown on the next page, but with additional elements up to element 120 not typically shown. Notice how the table appears more orderly with 120 elements, but Helium still looks misplaced once you consider the periodic table along with the four element blocks, s, d, p & f. It leaves the s-block incomplete, and the p-block with an extra element.

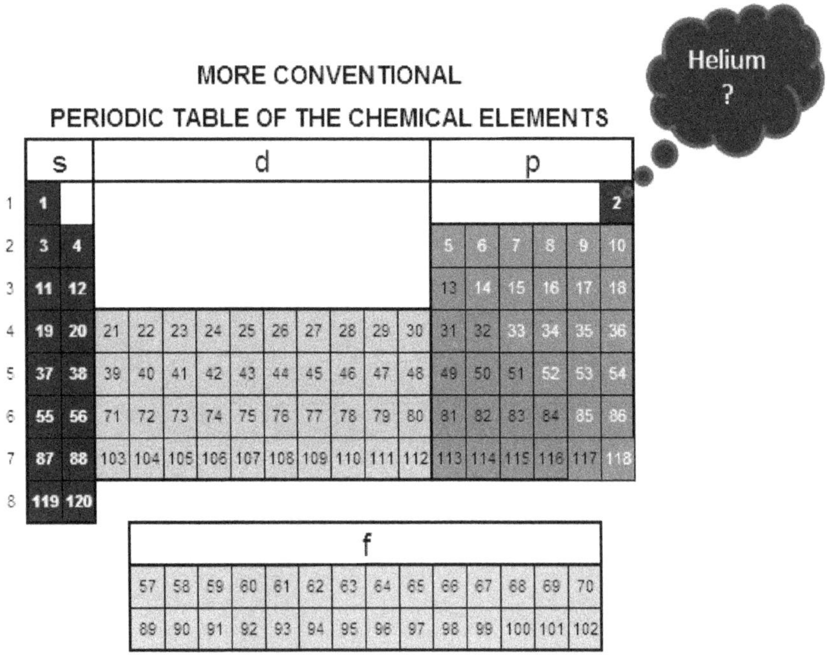

MORE CONVENTIONAL
PERIODIC TABLE OF THE CHEMICAL ELEMENTS
Helium
?
s
d
p
f
1 1 2
2 3 4 5 6 7 8 9 10
3 11 12 13 14 15 16 17 18
4 19 20 21 22 23 24 25 26 27 28 29 30 31 32 33 34 35 36
5 37 38 39 40 41 42 43 44 45 46 47 48 49 50 51 52 53 54
6 55 56 71 72 73 74 75 76 77 78 79 80 81 82 83 84 85 86
7 87 88 103 104 105 106 107 108 109 110 111 112 113 114 115 116 117 118
8 119 120
57 58 59 60 61 62 63 64 65 66 67 68 69 70
89 90 91 92 93 94 95 96 97 98 99 100 101 102

The conventional table of the chemical elements also leaves out the eighth period, which someday will be proven to exist when elements 119 and 120 are eventually discovered.

On the next page is yet another arrangement of the chemical elements that I believe Holy Spirit revealed to me by prompting me with the idea to center the four element blocks.

This arrangement leaves the table out of atomic number sequence, but it is interesting that this new arrangement forms an Isosceles-triangular shape, and when you add up the quantities of elements per row, a secondary sequence forms which looks *exactly* like the electron shell configuration of element 120. In fact, it is. The elements are in periodic order as well – eight periods (energy levels). The squiggly lines indicate the subshell fill sequence similar to Linus Pauling's subshell fill diagram turned upside down:

SPDF Table of the Elements (Isosceles-Triangular Version)

Based on the Electron Configuration Line Symmetry of Element 120, **(2, 8, 18, 32, 32, 18, 8, 2)**

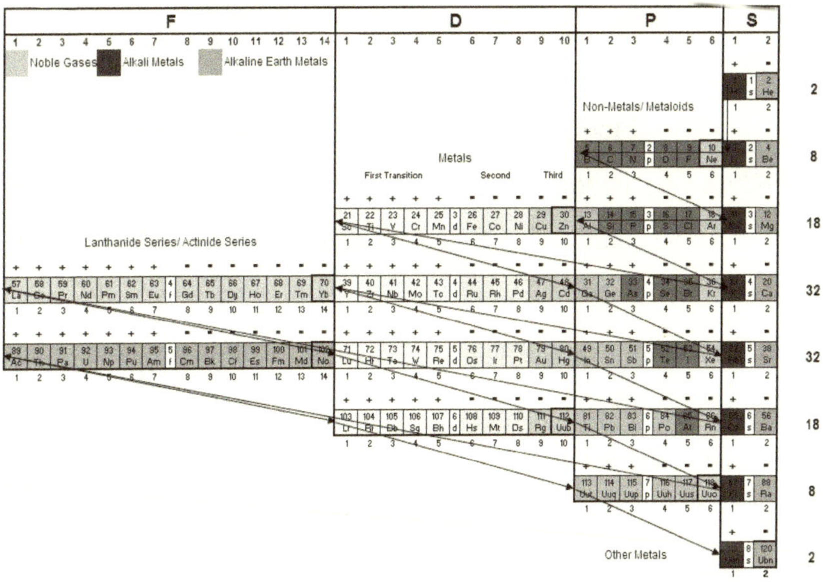

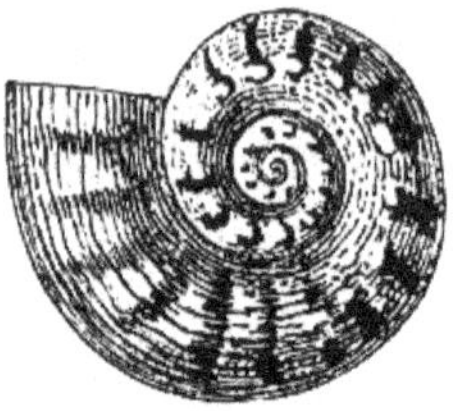 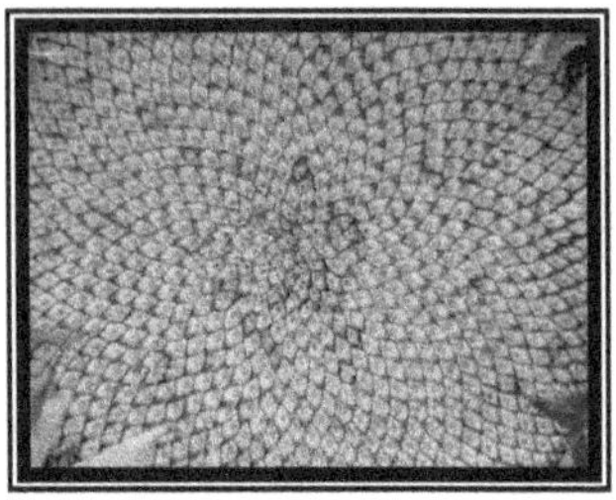 

Universal Patterns

You may already be aware of the existence of universal patterns. Throughout history there have been periods when what has been referred to as the Fibonacci sequence, the Golden Ratio or Phi, the Golden Mean, the Golden Angle, the Golden Rectangle, the Golden Triangle, and so forth, have become popular because of the many recurrences throughout the cosmos.

The Fibonacci sequence is a mathematical pattern produced by taking 0 and 1, 1 and 1, or 1 and 2, or any two consecutive numbers of the sequence, then adding them together to create the next number of the sequence. So, any number of the sequence is the sum of the previous two numbers. 0+1= 1; 1+1=2; 1+2=3; 2+3=5, 3+5=8, etc., producing the pattern: 0, 1, 1, 2, 3, 5, 8, 13, 21, 34, 55, 89, 144, and so forth. It is a pattern that can be observed throughout the universe, albeit seldom perfectly, just like line or mirror symmetry, where everything that exhibits line or mirror symmetry usually has a complete right and left side, even though both sides are seldom perfectly symmetrical. Yet when they appear more symmetrical than not, they are usually referred to as being symmetrical.

The same is true for the Golden Spiral found in the shapes of hurricanes from satellite views, or in the spirals of Nautilus shells, or in the spirals of distant galaxies; although not the perfect Golden Spirals mathematically, they certainly appear to be a close average of the spiral produced from the Fibonacci sequence which produces the ideal pattern.

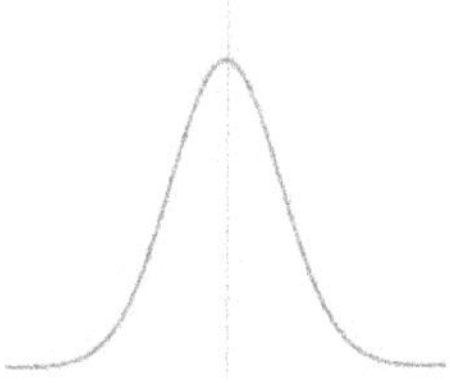

The Bell Curve is another example of a universal pattern. It is associated with the Law of Probability. It seldom is perfect, but it appears naturally nonetheless.

If you are unfamiliar with the Fibonacci sequence and want to know more about it, there are many websites that exploit these interesting phenomena. It is informative to conduct a general search on an Internet search engine, looking for "Fibonacci sequence." However, if you do not have much time to conduct a general search, you may want to take a look at these three websites:

1. "Fibonacci Number," on Wikipedia, the free encyclopedia online,[116] gives a good general description of what the Fibonacci sequence is about. Be sure to follow some of the referenced links.

2. "Fibonacci Flim-Flam," by Donald E. Simanek,[117] who takes each hyped up or exaggerated feature of the Fibonacci phenomena being generally broadcast on the Web, and attempts to set these exaggerated views straight. He tends to be somewhat condescending in the way he points out the faults of how the Fibonacci series has been mystified, hyped, and misrepresented by others, although in his effort to demonstrate the inaccuracy of specific illustrations, it seems to me he misses the main point of the whole thing—a point that the next recommended site makes clear.

3. "Brantacan," is the website of an individual whose area of expertise seems to be curved lines as related to bridges, arches in architecture, sound waves, and hyperbolic and sinusoidal angles. Brantacan,[118] from the UK, discusses and demonstrates with excellent

---

[116] http://en.wikipedia.org/wiki/Fibonacci_number
[117] http://www.lhup.edu/~dsimanek/pseudo/fibonacc.htm
[118] http://www.branta.connectfree.co.uk/fibonacci.htm

computer generated examples that can be seen and heard, the genuine relationships of the Fibonacci numbers, and resulting shapes related to those numbers to the shapes found in nature and in the universe.

Brantacan shows that the deviations of the individual samples are not so important as the net result of the many, or the bigger picture. Remember, most mirror symmetry is imperfect as well, but that doesn't discount the fact that nature functions magnificently having been adorned in its fabulous array of structured imperfection.

From a distance, all the daisies in the field look alike…but *are* they?

Part IV: Truth Science Philosophy

Chapter Fifteen

What Is Your Philosophy?

There is a prideful, atheistic bias that underpins today's science. You can see it in nearly every science magazine and publication. You can hear it in the science lectures from the elementary schools on up to the university levels. Its roots include the Descartes' perceived truth philosophy, which is essentially based on self-importance or pride, but it is also rooted in the Darwinian, Naturalist-Atheist influence as well. As a result, their refusal to acknowledge God as the Creator of all things has caused their vision to be reduced. In fact, it is this unbelief that has caused blind spots in their vision, and it has impaired their hearing so they cannot hear the still, small voice of God regarding matters of scientific significance.

The Bible speaks about those who are not guided by the truth, but who are guided instead by false doctrines, for they are the ones who are "always learning but never able to acknowledge the truth."[119] The Bible also says, "They will turn their ears away from the truth and turn aside to myths,"[120] and "They exchanged the truth of God for a lie, and worshiped and served created things rather than the Creator."[121]

Too many of today's professors teach these doctrines whether willingly or unknowingly as the result of having been indoctrinated themselves by the US Government education system, but these doctrines have become an integral part of the "Progressive" scientific methodology of today. There is indeed a philosophy to consider—it is our own.

As Christians, the knowledge of what truth is, and the careful application of truth-finding principles can be powerful tools for the study of science. Those willing to recognize and to reject the biases and doctrines that have been implanted via the public education system and many private institutions as well, and who are willing to instead seek and embrace the truth, will surely prove to be productive as they lay a stable foundation on which to construct their knowledge, understanding and wisdom; For if knowledge is not supported

---

[119] 2 Timothy 3:7 NIV
[120] 2 Timothy 4:4 NIV
[121] Romans 1:25 NIV

by truth, it will inevitably crumble into insignificance and ultimately translate into wasted time.

It is my personal opinion that time would be better spent *not* debating with evolutionists and Darwinian philosophers who perceive science through deaf ears and blind eyes. But instead, *Truth-Science* is dedicated and focused on the utilization of truth-finding principles, applied through a mind that understands the Biblical perspective, along with ears that are willing to hear the voice of God, and with eyes that are seeking to recognize God's signature and fingerprints in all of creation.

It is so amazing. There is such an enormous opportunity here for those who have eyes to see. There are many holes in science structures today due to the lack of willingness of others to perceive God in matters of science. The arrangement of the standard table of the chemical elements is only one example, with Helium having been misplaced, yet maintained by modern science. This therefore provides an enormous opportunity for discovery for those believers who are willing to approach science differently.

By utilizing the knowledge and understanding of God as the Creator that is included in a Biblical world view, and by incorporating a faith-based dynamic friend-relationship with

God into studies and research, believer scientists and science students have the potential and tremendous opportunity to discover things that have been previously overlooked by those who have approached science from the standard Descartes-Darwinian bent or perspective.

As believers acknowledge God as the Creator and designer of the universe– when they observe his signature in the world that surrounds them, and learn to recognize God's motifs and fingerprints—as they adopt and incorporate a philosophy of science based on truth into their everyday tasks, they will gain knowledge, understanding and wisdom of God's creation that will propel them forward. As they ask God specific questions, expecting to get answers and take the time to listen, they will hear God's still, small voice, and they will be taught and led by Holy Spirit as he reveals his handiwork to them in ways that unbelievers could never understand.

A well-founded philosophy based on truth can be a powerful tool to help a person maintain focus, productivity and effectiveness. A philosophy is a stronghold of thoughts, ideas and concepts that has been adopted into one's own heart and that remains anchored there by a person's agreement, acknowledgement and belief. It is an adopted set of principles. A well-founded philosophy of science based on an understanding and devotion to the truth will provide a stable foundation upon which knowledge, understanding and wisdom may be constructed. It will produce a good, positive bias that will quicken the vision and the hearing of the individual as he or she investigates, explores, observes, and conducts research and studies from a Biblical perspective.

On the other hand, having no particular philosophy or having a poorly founded one can lead to distraction, distortion, and wasted time. A poorly founded philosophy can lead one astray. It can produce a bad, negative bias that may even prevent the individual from seeing what is really there. Its effects can be similar to that of a cult, isolating the person from the truth.

So unless and until a person considers and adopts a well-founded philosophy based on an understanding and devotion to the truth, he or she has only what is there already, whether it is nothing (unlikely), or whether it is the subtle indoctrinations and intellectual programming of the public education system as well as other influences of the past, such as from friendships or acquaintances past and present.

Ask God to reprogram you. Ask him to make you aware of your biases, and then repent for agreeing with those beliefs and biases that are contrary to the Biblical worldview and with a faith-based, truth-based relationship with him. Then ask God to help you replace those bad biases with good ones.

You should always go to God first, for he is the source of all truth, and ask him in faith for wisdom about specific issues, believing that he will give you specific answers to your questions. It would be a good thing to write down a list of what you are asking God for, so that as you pray, you will be reminded to bring up all your questions and remain focused. This list can also serve as a check-off list, for as God reveals things by answering your questions, you can then check them off of your list. The list will then serve as a memorial and testimony to God's faithfulness in your life.

Learn to be comfortable talking with God. Use common language, and explain to him what you are doing and why. Then ask him your specific questions or make your specific request. After you have finished, thank him in advance by faith for the answers, and remember, you have just spoken with the God and Creator of the universe! Thank him, for who he is, and then ask Holy Spirit to lead and direct you in your study or investigation. The Bible says to "Trust in the Lord with all of your heart and lean not on your own understanding; in all of your ways acknowledge him, and he will direct your paths."[122] But remember, the Bible also says, "...You do not have, because you do not ask God."[123] So don't be afraid to ask him for anything, and never assume he will give you what you have not even asked him for.

After speaking with God, meditate for a few moments and allow God to speak to you. He will often, but not always, plant an idea in your mind. Run with it! You will understand the significance later on. Be flexible and open-minded. God often gives a picture or vision of something, or sometimes a single word, phrase or a quote, or a particular scripture will come to mind. Sometimes it will be a dream. Sometimes you will suddenly just know what to do. Sometimes you just have to begin in faith believing God is directing you, and he will—trust him, his timing is impeccable!

You should always consult the written Word of God for information regarding the subject you are studying or investigating. If you are not very familiar with the Bible, you may consider asking Christian friends who are more knowledgeable of the Bible than you, to see if they can recall any significant Bible passages or stories regarding or relating to the subject matter. Another excellent research tool, as discussed before, is using a Bible search engine such as

[122] Proverbs 3:5-6 NIV
[123] James 4:2

BibleGateway.com to conduct a word search or phrase search. The bottom line is that you must become familiar with God's word so that Holy Spirit can use it to direct you daily. So, frequent Bible study is always a concurrent course with all the others.

A God Bias

Unbelievers may think that consulting the Bible for information related to science is simply foolishness. Some think Christians have a religious bias that has no place in *real* science. Regardless of what others may think, having a God bias is a positive thing. There are bad biases and there are good biases.

I started high school P.E. in my freshman year with a coach that remembered one of my older brothers by name whom he had liked, and who had attended this coaches' P.E. class before me. Since my brother was good at sports, and because the coach liked my brother, he automatically liked me and probably assumed I would be good at sports as well. I *was* good at sports, but the coach's bias, thanks to my brother, helped him look for and recognize those traits in me faster than he would have had he never known my brother. So it was easy for me to do well in his class from the start.

A person with a God bias can recognize certain qualities or characteristics more quickly than others without it. This is because a person with a God bias is looking for and expecting to see specific qualities or characteristics. That is why having a God bias can be an advantage for a believer scientist. The person with the God bias will know what to be looking for, like a successful gold prospector who knows how and where to find gold.

Approaching science through a Biblically based faith perspective will enable a believer to see things that unbelievers have overlooked or misunderstood because they dismiss God as being the Creator of the universe. As believers, we truly can develop the ability to recognize God's fingerprints in creation because the rest of the scientific world trusts only in what it sees, hears, tastes, smells or feels—whereas Holy Spirit not only sharpens our senses, but also our minds based on our relationship with God.

How vs. Why

It takes more time, effort and resources to understand why something works than it does to understand how it works. It can be much more productive to focus and commit resources to understand how something works rather than to attempt to understand why it works, unless you are a designer. We seldom need the full spectrum of knowledge about something in order to use it. We need to come to an understanding or workable knowledge. We have PhD professors that exist for the hows and whys of things. Many spend a lifetime seeking the why knowledge of their specific field, when the majority of their students merely need to understand the hows. Within any subject there is a great deal of FYI knowledge. FYI knowledge is merely knowledge that is for your information only; it may be interesting, but it is not practical information to know.

For example, one does not need to know why a musical instrument works in order to play beautiful music. However, one must understand how it works, and must develop those essential skills to master it. In order to play trumpet beautifully, one must do exercises to train the fingers which keys to press down in order to play the different notes. One must do blowing and tonguing exercises that will train the ear and strengthen the lungs, lips and tongue in order to achieve pleasant-sounding tone and dynamics, and one must mostly practice.

Understanding how to operate a motor vehicle safely is essential if one wants to drive. One must know how to operate the steering wheel and the various pedals, levers, knobs and other controls. When driving where there are other vehicles on the road, one must also understand and obey the rules of driving or otherwise damage, injury or death may be the results. But understanding why a vehicle operates is not essential information unless you intend to design them.

Knowledge, Understanding & Wisdom

Knowledge, understanding and wisdom are not the same. Knowledge is information, and students generally find themselves bombarded with huge amounts of information. They quickly realize they will not be able to retain and assimilate all of the information they have received. That is why it is very beneficial to be trained in good note-taking, and in learning how to index information so that after they graduate they can review that information on an as-needed basis, and by doing so, develop a better understanding of the material later on.

Understanding is a higher form of knowledge, because it perceives the relationships and usefulness of that knowledge. With understanding comes responsibility– the responsibility to use that understanding for what it is worth, and not to waste it.

Knowledge is like the parts of a vehicle. Understanding is like the assembled vehicle. Okay, we know how the parts fit together and they are assembled, but the assembled vehicle is not enough to get you anywhere. To get somewhere, you need fuel for the vehicle. Wisdom is the fuel that will get you where you want to go. Wisdom is the skillful and strategic use of understanding. But how do you gain wisdom?

The Bible says, "For the Lord gives wisdom, and from his mouth come knowledge and understanding."[124] "The fear of the Lord is the beginning of wisdom, and knowledge of the Holy One is understanding."[125] "If any of you lacks wisdom, he should ask God, who gives generously to all without finding fault, and it will be given to him."[126] Why not take God at his word by asking him for knowledge, understanding and wisdom on a daily basis in order to maximize the time you have on this earth?

---

[124] Proverbs 2:6 NIV
[125] Proverbs 9:10 NIV
[126] James 1:5 NIV

Chapter Sixteen
A Journey Like None Other

So, for those of you who are willing to ask God specific questions, expecting to get specific answers, and who are willing to trust the guidance of Holy Spirit—to those of you who are willing to humble yourselves, and to publicly acknowledge God as the Creator of the universe, giving him the honor and recognition for those things he reveals to you—to all those who are willing to study and seek the truth about God's creation, and to learn how to recognize his fingerprints in the many patterns, cycles, layers, and perspectives of his handiwork throughout the universe and in the cosmos, from micro to macro and in between—to those of you who are willing to develop a God bias, growing in knowledge, attaining understanding based on truth, and seeking wisdom from on high—yes, to those of you who dare to be different in these ways, lies a vast treasure trove of discovery, opportunity and fulfillment.

Those who dare to be different will see things differently than the world sees things. Those who are not of this world will do things differently than the world does them.

Remember, it's not what you know, but whom you know that can make the biggest difference. Therefore, I challenge you to know God like you've never known him before, and to be dynamic in your relationship with him. God speaks any way he chooses. He is limitless.

If you seek, you *will* find. If you knock, the door *shall* be opened. So what are you waiting for? Meet *now* with the One Who knows you best.

Finally, may all of your endeavors be inspired and thrilling, and may your discoveries be awesome! Be blessed in all you become and in everything you do and accomplish. But mostly, be who God made *you* to be.

Never forget that you are artistically unique, and that God uses everyone uniquely whose heart's desire is to know him better – in an interpersonal way – not religiously. God has a special journey laid out just for you. It is a journey like none other. It is the experience of a lifetime, a journey that is yours alone. But remember, "God is a Spirit: and they that worship him must worship him in Spirit and in truth."[127]

[127] John 4:24 (KJV)

Appendix A

Word Usage Study

A Word Usage study can be easily performed with the use of an Internet Bible search tool. *BibleGateway.com* is an excellent example of a website that offers such a tool. It is available for anyone to use for free as long as it is for personal use, but I am sure they would appreciate a donation—it's well worth it for the time it can save you.

For example, by conducting a keyword search first for the word *soul*, and then for the word *spirit*, the Bible search engine will print to screen all the verse examples found throughout the entire Bible where the words *soul* and *spirit* were used, but you should search each word separately first. Then, by looking at all the examples where each word was used, you can begin to see the various ways each word was used. You may want to copy-and-paste several verses into a document, sorting them by meanings so as to collect several examples of each type of usage for comparison and further study later on.

For those who are familiar with the Bible, and have read it for many years, this process will generally take less time, because they are already familiar with much of the context surrounding the many scriptures that will appear.

You can further employ the use of other valuable Internet Bible search tools such as are offered at Heartlight's website, *SearchGodsWord.org.* Heartlight offers many Bible study tools, including an online Interlinear Bible that is very useful to help you find which Hebrew or Greek words were actually used in the original language. All you do is type the scripture reference into the search box. Then, it returns the verse in English along with the verse in either Hebrew or Greek beneath it, depending on whether the reference was from the Old or New Testament.

Each English, Greek, and Hebrew scripture verse displayed includes hyperlinks that will take you directly into a lexicon so you can see the source words, learn how those words were translated and used throughout the old or new testaments, and see how many times they were used for each translated meaning. You can also read the brief commentaries from Hebrew or Greek scholars to help you compare the source words with the English words they were actually translated into.

www.ingramcontent.com/pod-product-compliance
Lightning Source LLC
Chambersburg PA
CBHW031339040426
42443CB00006B/393
* 9 7 8 0 6 1 5 4 3 6 5 1 7 *